Environmental Information for Naval Warfare

Committee on Environmental Information for Naval Use

Ocean Studies Board

Division on Earth and Life Studies

NATIONAL RESEARCH COUNCIL
OF THE NATIONAL ACADEMIES

THE NATIONAL ACADEMIES PRESS
Washington, D.C.
www.nap.edu

THE NATIONAL ACADEMIES PRESS • 500 Fifth Street, N.W. • Washington, DC 20001

NOTICE: The project that is the subject of this report was approved by the Governing Board of the National Research Council, whose members are drawn from the councils of the National Academy of Sciences, the National Academy of Engineering, and the Institute of Medicine. The members of the committee responsible for the report were chosen for their special competences and with regard for appropriate balance.

This report and the committee were supported by a grant from the Office of Naval Research and the Office of the Oceanographer of the Navy. Any opinions, findings, conclusions, or recommendations expressed in this publication are those of the author(s) and do not necessarily reflect the views of the organizations or agencies that provided support for the project.

International Standard Book Number 0-309-08860-7

Additional copies of this report are available from the National Academies Press, 500 Fifth Street, N.W., Lockbox 285, Washington, DC 20055; (800) 624-6242 or (202) 334-3313 (in the Washington metropolitan area); Internet, http://www.nap.edu

Copyright 2003 by the National Academy of Sciences. All rights reserved.

Printed in the United States of America

THE NATIONAL ACADEMIES
Advisers to the Nation on Science, Engineering, and Medicine

The **National Academy of Sciences** is a private, nonprofit, self-perpetuating society of distinguished scholars engaged in scientific and engineering research, dedicated to the furtherance of science and technology and to their use for the general welfare. Upon the authority of the charter granted to it by the Congress in 1863, the Academy has a mandate that requires it to advise the federal government on scientific and technical matters. Dr. Bruce M. Alberts is president of the National Academy of Sciences.

The **National Academy of Engineering** was established in 1964, under the charter of the National Academy of Sciences, as a parallel organization of outstanding engineers. It is autonomous in its administration and in the selection of its members, sharing with the National Academy of Sciences the responsibility for advising the federal government. The National Academy of Engineering also sponsors engineering programs aimed at meeting national needs, encourages education and research, and recognizes the superior achievements of engineers. Dr. Wm. A. Wulf is president of the National Academy of Engineering.

The **Institute of Medicine** was established in 1970 by the National Academy of Sciences to secure the services of eminent members of appropriate professions in the examination of policy matters pertaining to the health of the public. The Institute acts under the responsibility given to the National Academy of Sciences by its congressional charter to be an adviser to the federal government and, upon its own initiative, to identify issues of medical care, research, and education. Dr. Harvey V. Fineberg is president of the Institute of Medicine.

The **National Research Council** was organized by the National Academy of Sciences in 1916 to associate the broad community of science and technology with the Academy's purposes of furthering knowledge and advising the federal government. Functioning in accordance with general policies determined by the Academy, the Council has become the principal operating agency of both the National Academy of Sciences and the National Academy of Engineering in providing services to the government, the public, and the scientific and engineering communities. The Council is administered jointly by both Academies and the Institute of Medicine. Dr. Bruce M. Alberts and Dr. Wm. A. Wulf are chair and vice chair, respectively, of the National Research Council.

www.national-academies.org

COMMITTEE ON ENVIRONMENTAL INFORMATION FOR NAVAL USE

PAUL E. TOBIN (*Chair*), Armed Forces Communications and Electronics Association, Fairfax, Virginia
THOMAS P. ACKERMAN, Pacific Northwest National Laboratory, Richland, Washington
ARTHUR B. BAGGEROER, Massachusetts Institute of Technology, Cambridge
E. ANN BERMAN, Tri-Space, Inc., McLean, Virginia
STEPHEN K. BOSS, University of Arkansas, Fayetteville
TONY F. CLARK, North Carolina State University, Raleigh
PETER C. CORNILLON, University of Rhode Island, Narragansett
CARL A. FRIEHE, University of California, Irvine
EILEEN E. HOFMANN, Old Dominion University, Norfolk, Virginia
ROBERT A. HOLMAN, Oregon State University, Corvallis
GAIL C. KINEKE, Boston College, Chestnut Hill, Massachusetts
JOHN M. RUDDY, Missile Defense Agency, Washington, DC

Staff

DAN WALKER, Study Director
JOHN DANDELSKI, Research Assistant
DENISE GREENE, Senior Project Assistant

OCEAN STUDIES BOARD

NANCY RABALAIS (*Chair*), Louisiana Universities Marine Consortium, Chauvin
ARTHUR BAGGEROER, Massachusetts Institute of Technology, Cambridge
JAMES COLEMAN, Louisiana State University, Baton Rouge
LARRY CROWDER, Duke University, Beaufort, North Carolina
RICHARD B. DERISO, Inter-American Tropical Tuna Commission, La Jolla, California
ROBERT DITTON, Texas A&M University
EARL DOYLE, Shell Oil (ret.), Sugar Land, Texas
ROBERT DUCE, Texas A&M University, College Station
WAYNE R. GEYER, Woods Hole Oceanographic Institution, Woods Hole, Massachusetts
STANLEY R. HART, Woods Hole Oceanographic Institution, Woods Hole, Massachusetts
MIRIAM KASTNER, Scripps Institution of Oceanography, La Jolla, California
RALPH S. LEWIS, Connecticut Geological Survey, Hartford
WILLIAM MARCUSON, U.S. Army Corps of Engineers, (Ret.)
JULIAN P. MCCREARY, JR., University of Hawaii, Honolulu
JACQUELINE MICHEL, Research Planning, Inc., Columbus, South Carolina
SCOTT NIXON, University of Rhode Island, Narragansett
SHIRLEY POMPONI, Harbor Branch Oceanographic Institute, Ft. Pierce, Florida
FRED SPIESS, Scripps Institution of Oceanography, La Jolla, California
JON G. SUTINEN, University of Rhode Island, Kingston
NANCY TARGETT, University of Delaware, Lewes

Staff

MORGAN GOPNIK, Director
SUSAN ROBERTS, Senior Program Officer
DAN WALKER, Senior Program Officer
JOANNE BINTZ, Program Officer
JENNIFER MERRILL, Program Officer
TERRY SCHAEFER, Program Officer
ROBIN MORRIS, Financial Officer
JOHN DANDELSKI, Research Associate
SHIREL SMITH, Administrative Associate
JODI BACHIM, Senior Project Assistant
NANCY CAPUTO, Senior Project Assistant
DENISE GREENE, Senior Project Assistant
SARAH CAPOTE, Project Assistant
BYRON MASON, Project Assistant
JULIE PULLEY, Project Assistant

Preface

During my years of naval service, I depended on accurate and timely information about atmospheric and oceanographic conditions derived from a limited number of sources. Today's operational commander can access multiple sources via high-bandwidth data paths. This plethora of information, however, can rapidly overwhelm his or her ability to incorporate that information into real-time decisionmaking. Fortunately, the tools are now at hand to evaluate the uncertainty associated with various information products, greatly facilitating decisions based on an ever-increasing volume of information.

Introducing 13 talented scientists to the meteorological and oceanographic (METOC) community and the challenges facing it was a pleasure. After many years of Naval Service, I am very proud of the organization and particularly of the METOC community where I spent my last two years of service. U.S. Naval Forces consist of two very large, multifaceted, complex organizations (i.e., the U.S. Navy and Marine Corps). Fortunately, scientists deal with complex systems routinely, and their ability to assimilate the details of naval METOC has been remarkable. The learning process involved climbing through a Nuclear Aircraft Carrier and a Guided Missile Destroyer and visiting major METOC activities on both coasts. I am indebted to the panel members for the generous use of their time and to the U.S. Navy and Marine Corps for opening every door we requested to pass through. Seeing the U.S. Navy and Marine Corps through the eyes of scientists was very useful and enlightening for me.

The U.S. Navy and Marine Corps have each made a great investment in their METOC personnel. This community is one of the most highly educated groups in either service. A comparable investment in collection platforms, sensors, computer models, and expendable resources to cover the entire world has not and cannot be made. Commanders, Pilots and Ship Captains all desire perfect METOC information, all of the time. There are tools now available that bring us as close to that goal as practical, provided all the needed resources are available at the desired location.

How close do we come to the goal of forecasting certainty? The answer is the traditional "it depends." If the location were Moorhead City, North Carolina, or Camp Pendleton, California, the degree of certainty would be high in the short term. We have studied these areas intensely over the last 75 years and hold many exercises at each yearly. Sensing resources are always available and climatological data are extensive, but even in these ideal cases we cannot adequately forecast parameters like wave height and coastal currents beyond 72 hours. Locations like Kandahar, Afghanistan, present a whole range of new problems, including limited access, sparse historical data, limited remote sensing, and a hostile climate. Another complicating aspect is the varied nature of naval missions. METOC information support for peacetime naval presence varies dramatically from the most difficult scenario of all, an amphibious operation. Finally, the enemy threat must be known, and we rely heavily on the Naval Intelligence Community to provide enemy intentions and weapons capabilities. Mission, location, season, friendly weapons choices, and enemy intentions are all part of a complex matrix that the Commander and METOC planners must confront. The degree of forecasting success thus depends on how well scarce resources are allocated across this broad range of factors.

Some requirements, such as safety in the air and on the sea, have been well supported over the years and although important do not offer the potential for large new payoffs through increased investments. Properly chosen increased support of warfighting mission areas could yield major gains in terms of weapons performance and ultimate victory. An example would be enhanced remote sensing through aircraft, autonomous aviation vehicles, or satellites in support of Precision-Guided Munitions. Reducing uncertainty here could be critically important. Current trends in weapons deploymneet suggest that this is and will continue to be the case.

Since we are dealing with scarce resources, it is no surprise that uncertainty and business models are areas that this study has found as keys to the future. We are very fortunate to have the tools at hand to ensure accurate forecasting and success in combat. We currently excel in this process, and if we make the right choices in the future, environmental uncertainty will not be completely eliminated but will be a far more manageable concern for our commanders. The following study will describe a process that I firmly believe will take us to that goal.

> Paul E. Tobin, RADM USN (ret.)
> *Chair,* Committee on Environmental Information for Naval Use

Acknowledgments

This report was greatly enhanced by the participants of the multiple information-gathering activities held as part of this study. The committee would first like to acknowledge the efforts of those who gave presentations at meetings. The following individuals provided significant insight by making formal presentations to the committee.

JOE ATANGAN
RUSS BEARD
JERRY BIRD
JERRY BOATMAN
MARK BOSTON
MELBOURNE G. BRISCOE
HOUSTON COSTOLO
DOUG CRONIN
TOM CUFF
JOHN GARSTKA
JERRY GATHOF
ALFRED GENT
CHRISTINE JARET
ROB LAWSON
STEVE LINGSCH
MICHAEL S. LOESCHER
STEPHEN MARTIN
DINTY MUSK
TERRY PALUSZKIEWICZ

JAMES RIGNEY
RICHARD SPINRAD
JOSEPH SWAYKOS
VAN GURLEY
PHIL VINSON

The committee also met with various Navy and Marine Corps personnel during seven subgroup meetings. These meetings were invaluable, and we would like to express our appreciation to each individual, but there are too many names to mention. The committee is also grateful to the Navy panel that provided important discussion and/or material for this report:

CAPT TY ALDINGER, Commander-in-Chief, Pacific Fleet; **LCDR JIM BERDEQUEZ,** DCNO for Expeditionary Warfare (N75); **LT COL (RET) TOM CUMMINS,** USCM, Defense Intelligence Agency; **DR. RON FEREK,** Office of Naval Research; **CAPT CHRIS GUNDERSON,** Deputy Oceanographer of the Navy; **LCDR VAN GURLEY,** Commander, Cruiser-Destroyer Group 2; **LCDR PAUL MATTHEWS,** Commander, Naval METOC Command; **LCDR TONY NEGRON,** Commander, Naval METOC Command; **DR. TERRY PALUSZKIEWICZ,** Office of Naval Research; **DR. RUTH PRELLER,** Naval Research Laboratory; **CAPT DAVE TITLEY,** Dept Asst. SECNAV (Mine/Undersea Warfare); and **CDR ZDENKA WILLIS,** Commanding Officer, Naval Ice Center.

The committee also owes significant thanks to the members and staff of the Naval Studies Board. The willingness of Alan Berman and Richard Ivanetich to review the report, of Art Baggeroer to serve on the committee, and of the NSB staff to provide assistance where possible reflects a genuine and deep commitment to helping the U.S. Navy and Marine Corps benefit from the best scientific and technical advice available.

This report has been reviewed in draft form by individuals chosen for their diverse perspectives and technical expertise, in accordance with procedures approved by the National Research Council's Report Review Committee. The purpose of this independent review is to provide candid and critical comments that will assist the institution in making its published report as sound as possible and to ensure that the report meets institutional standards for objectivity, evidence, and responsiveness to the study charge. The review comments and draft manuscript remain confidential to protect the integrity of the deliberative process. We wish to thank the following individuals for their participation in the review of this report:

ACKNOWLEDGMENTS

DR. ALAN BERMAN, Independent Consultant, Alexandria, Virginia
REAR ADM MILLARD S. FIREBAUGH, General Dynamics, Electric Boat Corporation, Groton, Connecticut
DR. DONALD P. GAVER, Naval Postgraduate School, Monterey, California
DR. RICHARD IVANETICH, Institute for Defense Analyses, Alexandria, Virginia
DR. ALFRED KAUFMAN, Institute for Defense Analyses, Alexandria, Virginia
DR. JOAN OLTMAN-SHAY, Northwest Research Associates, Inc., Bellevue, Washington
REAR ADM RICHARD F. PITTENGER, Woods Hole Oceanographic Institution, Massachusetts
DR. ROBERT WELLER, Woods Hole Oceanographic Institution, Massachusetts
GEN KEITH SMITH, U.S Marine Corps. (ret.), Vienna, Virginia

Although the reviewers listed above have provided many constructive comments and suggestions, they were not asked to endorse the conclusions or recommendations nor did they see the final draft of the report before its release. The review of this report was overseen by Brad Mooney, appointed by the Divison on Earth and Life Studies, who was responsible for making certain that an independent examination of this report was carried out in accordance with institutional procedures and that all review comments were carefully considered. Responsibility for the final content of this report rests entirely with the authoring committee and the institution.

Contents

EXECUTIVE SUMMARY		1
1	INTRODUCTION AND OVERVIEW	12
2	THE VALUE OF ENVIRONMENTAL INFORMATION	34
3	NATURE OF THE PROBLEM: SOURCES AND LIMITATIONS OF METOC KNOWLEDGE	53
4	IMPROVING ENVIRONMENTAL INFORMATION BY REDUCING UNCERTAINTY	70
5	INFORMATION FLOW: LEVERAGING NETWORK-CENTRIC CONCEPTS	110
6	MOVING AHEAD	124
REFERENCES		138
APPENDIXES		
A	COMMITTEE AND STAFF BIOGRAPHIES	143
B	THE ROLE OF ENVIRONMENTAL INFORMATION IN NAVAL WARFARE	147

C	ENVIRONMENTAL SCIENCE AND TECHNOLOGY PROGRAMS	172
D	ACRONYMS	195
E	INFORMATION-GATHERING ACTIVITIES OF THE COMMITTEE ON ENVIRONMENTAL INFORMATION FOR NAVAL USE	201

Executive Summary

Understanding the distribution of friendly, enemy, and neutral forces and facilities and the nature and significance of the environment they occupy is key to what the Joint Chiefs of Staff described as battlespace awareness in *Joint Vision 2020*.[1] A shared awareness of the battlespace among allied military forces is considered to be a force multiplier that is recognized as both highly desirable and difficult to achieve. Before data can be meaningfully shared, data must first be obtained and then rendered into a usable form—in other words, information or knowledge. Efforts to collect, assimilate, analyze, and disseminate information about and predict the nature, distribution, and intent of enemy forces have long been the focus of a large and complex intelligence, surveillance, and reconnaissance effort. Efforts to collect, assimilate, analyze, and disseminate information about and predict the nature and significance of the environmental character of the naval battlespace, though less well known, have been the focus of a complex meteorological and oceanographic effort referred to within the U.S. Naval Forces as METOC.

There are striking similarities between intelligence, surveillance, and reconnaissance data gathering and METOC data-gathering efforts. For example, the availability and accuracy of information provided to military decisionmakers in

---

[1]*Joint Vision 2020* (JV 2020) is the Chairman of the Joint Chiefs of Staff's vision for how America's Armed Forces will transform in order to create a joint force that is dominant across the full range of military operations. It calls for achieving full spectrum dominance focused on four operational concepts: dominant maneuver, precision engagement, focused logistics, and full-dimensional protection as enabled by information superiority, innovation, and increased joint, interagency, and multinational interoperability. January 31, 2003 (http://www.dtic.mil/jv2020/jvpub2.htm).

the past have been uneven, causing leaders at various levels to view its value with some skepticism. In many situations, military strategists and field commanders "planned for the worst" and concentrated as many resources as possible on achieving an objective, in order to overcome uncertainty through overwhelming force. In the geopolitical landscape of the 21st century, "blunt force" warfighting tactics have necessarily and thankfully been augmented, if not replaced, by what in contrast are surgically precise strikes. The longstanding need to minimize casualties among friendly forces and noncombatants has, in turn, dramatically reduced the margin for error. Under such circumstances, the need for accurate, reliable, and timely intelligence, surveillance and reconnaissance, and related environmental information (i.e., total battlespace awareness) is even greater. Although few military planners would trade knowledge about the distribution and intent of enemy forces for improved environmental understanding, failure to acknowledge the importance of environmental conditions has played a role in many failed military operations that were otherwise well planned and executed.

To aid in the development of an investment strategy to enhance the value of METOC contributions to battlespace awareness, the Office of the Oceanographer of the Navy and the Office of Naval Research (ONR) requested that the National Academies undertake a two-year effort to:

• develop a framework process that can be adapted by U.S. Naval Forces to prioritize what data should be collected and how the data should be managed, what new models should be developed and what improvements should be made to existing models, which data fusion and value-added products should be developed to effectively and efficiently disseminate environmental information to naval units;

• identify those segments of the process that would benefit from targeted research (e.g., specific ocean processes or general areas of uncertainty); and

• prioritize the proposed improvements by identifying which actions are most needed and achievable as well as those that are most likely to make a significant and positive impact.

Although the similarities between intelligence gathering, surveillance and reconnaissance, and environmental characterization are numerous and the opportunities for synergy are significant, the focus of this report is on how best to enhance the environmental information component of naval battlespace awareness. Over the past 300 years considerable effort has gone into understanding and characterizing the naval environment. The emphasis has been on reducing uncertainty about the conditions that naval forces face, since it is uncertainty that affects planning for desirable outcomes. In the most recent decades the development and implementation of complex weapons systems have largely been driven an increased need for precise environmental information. Although all-weather

EXECUTIVE SUMMARY

weapons systems are being developed, understanding the effect of any variety of environmental parameters on weapons system and platform performance will remain an important component of tactical decisionmaking for the foreseeable future.

The current METOC enterprise and its predecessor organizations have brought the U.S. Naval Forces high-quality environmental information that has served them well in peace and in war. However, as the Department of Defense transforms its force structure to meet the challenges that now face the nation, METOC must also examine how it will support the future. New approaches will be needed to provide METOC customers with information and knowledge more rapidly, anywhere, and at any time. This will require new ways to collect the necessary data, new ways to analyze those data to create and present information, and new ways to deliver or make available that information worldwide to advantaged and disadvantaged users alike.

Many previous efforts by the National Research Council focused on specific environmental processes of concern to naval forces or advocated specific actions to better understand or characterize those processes. While some discussion of those issues is appropriate, this report focuses more extensively on underlying the philosophies needed to determine when and how to improve environmental information. Thus, understanding what effect uncertainty in environmental information has on the ability of U.S. Naval Forces to execute critical missions, what the cost of that uncertainty is, and identifying what efforts might reduce these emerged as a significant unifying theme in this report. In a very real sense, understanding when environmental knowledge is central to mission success, how "certain" that knowledge needs to be (and at what temporal and spatial scales) defines the business model that will serve to describe the framework for investment decisions. U.S. Naval Forces often have an intuitive feel for when environmental information is important to improving the likelihood of mission success. More difficult to quantify are the levels of uncertainty that transform missions with high likelihood of success into marginally successful or failed missions. Likewise, it is difficult to quantify an acceptable cost for reducing the uncertainty in environmental predictions.

The major findings and recommendations in this report are based on a business model approach to the METOC mission areas that encompasses allocation of investment and management of risk in order to:

• ensure that environmental information is presented in a manner that conveys to the warfighter an appropriate level of confidence in its content,
• ensure that efforts to enhance environmental information are carried out in a cost-effective manner that maximizes resources, and
• create a system for information exchange that allows a high degree of informed involvement by warfighters.

DEVELOPING MORE EFFECTIVE METOC PROCESSES: NEAR-TERM LEVERAGING OF EXISTING PROGRAMS AND RESOURCES

Information about environmental conditions is developed from observation or inference. Understanding the nature of environmental processes provides the skill needed to determine when additional observations are needed to interpret past conditions to predict present conditions at a location of interest (e.g., provide information on beach trafficability for a denied coastline) or to build on existing information to draw inferences about conditions in the future (e.g., forecast weather over a carrier battlegroup three days in the future). The latter process is at the heart of traditional forecasting and has been greatly expanded by advances in the capability and capacity of computing facilities and remote sensing. Much work, however, remains to be accomplished if forecasting is to achieve the accuracy and reliability needed at the temporal and spatial scales relevant to many naval operations, especially those taking place in coastal areas.

In the presence of perfect battlespace awareness, perfect tactical decisions are theoretically possible. Perfect environmental information, however, is neither achievable or even necessary in many instances. Furthermore, it will not be possible to obtain, manage, and disseminate environmental information at all scales of interest for all areas of possible naval activities in the foreseeable future. Like many entities with an operational focus, the naval METOC enterprise has evolved to operate on very short production cycles. At present, different information for various geographic areas of concern is distributed over multiple sources, many of which are identified by various METOC officers "in stride" as reports are developed for various customers in response to real-time requests. Success in this approach is largely dependent on the knowledge, creativity, and experience of individual METOC officers. Lack of a more cohesive or proactive approach to priority setting limits the METOC community's ability to identify, evaluate, and acquire data and information from nontraditional sources during emerging crises. The METOC community needs to become both a supporter of network-centric operations and a beneficiary of those operations by being an active user of the networks being developed to support them. **The METOC enterprise should incorporate network-centric approaches to enable easy and flexible interconnectivity among the individual METOC officers and with nontraditional sources of information. This should be done now to leverage the people and knowledge assets currently in place.**

Special attention should be given to identifying the METOC contribution for nonroutine operations (e.g., evacuation of noncombatants, amphibious warfare, as opposed to activities such as ship tracking or air operations that occur on a daily basis and thus tend to be continuously evaluated and modified). Guidance for identifying broadly needed and significant information across multiple warfare areas should be derived from an understanding of the benefit of additional information for reducing uncertainty versus the cost of improving the content and reliability of environmental information, whether through additional observations, improved understanding of the underlying physical processes, or more powerful forecasting tools that take advantage of both.

The availability of unmanned airborne vehicles (UAVs) and unmanned underwater vehicles, and their expanded capability to covertly collect intelligence, surveillance, and reconnaissance information in denied areas using a variety of electrooptical and acoustic sensors, creates a largely untapped potential for the unintended use of such information to support the development or validation of METOC products or forecasts. **The Oceanographer of the Navy and the Commander of Naval Meteorology and Oceanography Command (CNMOC) should work with the broader community within the U.S. Department of Defense (DOD) and elsewhere to expand efforts to make intelligence, surveillance, and reconnaissance information and data with environmental content more accessible to the METOC community.**

Expanded efforts should be included to remove unneeded or particularly sensitive nonenvironmental content, thus reducing security risk while making the environmentally relevant information or data acquired during intelligence-gathering, surveillance, and reconnaissance efforts more accessible to the METOC community. At the same time, the METOC community's ability to securely handle sensitive georeferenced material must be expanded. **The METOC community should also seek to acquire its own unmanned platforms since concerns about security limit the availability of such collateral information. In addition, it is strongly recommended that ONR evaluate the potential for exploiting existing intelligence-gathering, surveillance, and reconnaissance sensors as dual-use METOC sensors. Assuming this potential is significant, the Oceanographer of the Navy and CNMOC should work with the DOD community to develop mechanisms to exploit this potential.**

Particular emphasis should be placed on forward-deployed assets that are already under the control of theater commands. Data collection could be either specifically tasked or carried out while en route to other missions. Care must be taken that the primary missions of the intelligence, reconnaissance, or surveillance operations involved are not hindered or compromised, so that such dual-use activities are indeed cost effective.

As network-centric warfare (NCW) becomes an operational reality, enhanced computing and communications capabilities are changing the way the U.S. Naval Forces fight, communicate, and plan. Extensive e-mail and classified METOC electronic chat room traffic is already overtaking the formal Naval Message System and creating peer-to-peer linkages that are a radical departure from the hierarchical system that has been in place for years. Faster computers and high-speed data links resulting from the IT-21 initiative and the Navy-Marine Corps Internet program are accelerating this dramatic change in the METOC community.

The Office of the Oceanographer of the Navy and CNMOC should work with regional METOC commands to formally define a network-centric concept of operations that embraces peer-to-peer networking within the METOC community while preserving the security, flexibility, and timeliness that have led to the rapid growth in its use. The goal of formalizing this type of exchange should be to improve information content and its usefulness as a source of insight into user and customer needs (e.g., opportunities for data mining, frequency of various types of information requests, identification of systematic problems in information access) while encouraging continued and wider usage. Access to, and the transmission of, METOC data between ships and to shore facilities will be significant parts of the NCW transformation. Current efforts to incorporate network-centric principles into METOC operational concepts are only beginning to tap the vast potential that the network-centric operational concept offers.

EXPANDING METOC CAPABILITIES: LOGICAL NEXT STEPS

Existing capabilities for data collection, storage, and dissemination can produce voluminous bodies of information with varying amounts of useful content. The sheer volume of available information is already posing an unforeseen challenge as the METOC community and the warfighters they support struggle to match useful information to key users. **The Oceanographer of the Navy should ensure that CNMOC, the Fleet Numerical Meteorology and Oceanography Command (FNMOC), and the Naval Oceanographic Office (NAVOCEANO) jointly develop a strategic plan for data acquisition over the next 10 to 20 years that prioritizes geographic regions of focus; incorporates an understanding of the limits of environmental information currently available to the METOC community for various regions; and evaluates such technologies as distributed databases, advanced information data-mining techniques, and intelligent agent technology for enhancing environmental information for priority areas.**

In addition to geopolitical considerations (which may be fairly fluid on decadal timescales), such a plan should be based on a thorough understanding of what environmental information is currently held or available and which processes or conditions will be of particular concern to various naval missions. **Once an initial framework is established, CNMOC and NAVOCEANO should work with operational commanders to evaluate the adequacy of existing critical information (e.g., external variables such as bathymetry/topography, sediment type/land cover) and plan for filling data and information gaps.**

Increasing bandwidth, while relaxing some constraints, will undoubtedly lead to further dilution of information content. As the locations of naval actions vary across multiple continents and adjoining seas, greater effort must go into developing mechanisms for rapid and efficient environmental characterization that focus on providing the warfighter with targeted information with a high proportion of useful content while minimizing ancillary or irrelevant information. High-quality data are imperative precursors of high-quality information. **The Office of the Oceanographer of the Navy should foster efforts by ONR (by providing expertise and access) to develop and implement a system that promotes Optimized Environmental Characterization, keyed to action-specific warfighter needs during various naval missions or suites of missions. Data and information quality are central.**

A significant component of the METOC enterprise, in terms of both fiscal and human resources, is devoted to data collection. Understanding how new data collection, as opposed to use of archived data or numerical extrapolation or interpolations, improves the content of environmental information (reduces uncertainty) should be a key component of targeted data acquisition. Since data collection resources are limited and because the cost of data acquisition in denied areas can be very high, methods for establishing data collection efforts and the research and development that support data collection platform development should be focused using objective criteria. **The Office of the Oceanographer of the Navy and CNMOC should invest in the development of formal and rigorous methods for identifying high-priority data needs that are specific to the platforms and missions to be involved in any potential naval action. Determination of the cost of uncertainty, and a focus on data collection efforts that result in the greatest reduction in total cost of uncertainty, should be given priority at ONR and the Office of the Oceanographer.**

Asset allocation should be based on achieving improvements in the most significant parameters of interest. Such an effort will need to be based on a rigorous understanding of critical thresholds for platforms and systems involved in, as well as the spatial and temporal variability of key parameters and the operational tempo associated with, each mission or suite of missions.

At present, there is insufficient continuity of responsibility and feedback for maintaining databases and models. In addition, there is inadequate exchange of data or information collected or managed throughout the Department of the Navy. **The Oceanographer of the Navy should clarify the various areas of responsibility and assess the performance of such databases and models, and the tactical decision aids that rely on them, focusing on their value for individual mission areas.** In other words, the value or adequacy of a specific database or model may vary by mission or project, but information collected for a specific mission or project may still be of value to unintended users. Furthermore, at present, there is insufficient use of datasets collected by other federal agencies and academia. **CNMOC and FNMOC should expand efforts to identify data of value and work with ONR to develop methodologies for evaluating those data and bringing them into existing METOC systems.** Once expanded capabilities to access data and information from a variety of sources, whether from within DOD, academia, or other nongovernment sources, is established, an effort should be made to develop and implement a system that permits rapid retrieval of environmental data collected in specific geographical areas.

The U.S. Marine Corps' traditional role in expeditionary warfare and renewed focus on littoral operations involving the U.S. Navy continue to drive the need for environmental information in coastal areas where access is frequently denied.[2] Efforts to improve secure, low-profile communications, reduce the risk to personnel in coastal areas from chemical and biological agents (either from the tactical deployment of weapons of mass destruction by enemy forces or the destruction of such weapons by friendly forces), and provide accurate assessments of atmospheric conditions during the planning and implementation of strike missions have placed a greater emphasis on coastal meteorology. This includes development of predictive meteorological capability at fine scales and intensive data gathering in coastal environments. Military operations in these areas require data in forward operations areas. UAVs are ideally suited for this and provide additional opportunity for data gathering on behalf of METOC. **The Office of the Oceanographer of the Navy should work with the Space and Naval Warfare Systems Command (SPAWAR) and the operational commands to further develop and deploy atmospheric sensors on UAVs that will permit the collection of essential environmental information without impairing the intelligence, reconnaissance, or surveillance efforts they are largely designed to carry out.**

Current efforts to model many littoral processes of importance are promising but not fully operational. Thus, needed predictive capabilities are not currently available. **The Office of the Oceanographer should foster efforts at ONR and fund efforts at SPAWAR to integrate mesoscale models with local littoral models.**

CHANGING ATTITUDES AND APPROACHES: A LONGER-TERM VISION

The philosophy and approaches used to supply METOC information to the fleet and Marine Corps can be described in terms of a business model (in fact, the U.S. Navy and Marine Corps, like the DOD and most federal agencies, have

---

[2]The geographic range over which the U.S. Marine Corps operates may extend landward many hundred of miles. The farther inland the Marine Corps operates, the less it relies on METOC information. For example, terrain analysis, land cover, and other factors affecting target acquisition are derived from information provided by many sources, including the National Imagery and Mapping Agency. Under such circumstances meteorological information may be provided by non-Navy DOD sources such as the U.S. Air Force or even coalition partners. Given the resources available, the Committee decided to limit its detailed discussion to environmental information provided by naval sources. (Reducing uncertainty or understanding how uncertainty limits the value of environmental information are issues of broad applicability. This study is intended to provide advice to the Office of the Oceanographer of the Navy and the Office of Naval Research.)

already adopted this philosophy to some degree, as testified by the widespread use of such terms as "user driven" and "customer needs" in planning documents). The present METOC business model[3] for providing global and mesoscale forecast fields is well defined and successful. However, the METOC business model for providing enhanced warfighting capabilities does not appear to adequately address customer needs. The current relationship does not facilitate close connectivity or fluid information exchange with the customer. Therefore, it is difficult to rigorously define the contribution of METOC to or its impact on improved warfighting.

It is unclear whether the existing METOC enterprise reflects a strategic or unifying principle that can help the various component parts understand their relationship to one another and the overall goal of the activity. Providing specific advice for improving the overall performance of the METOC enterprise is therefore difficult. A review of various DOD and Department of the Navy guidance documents, however, does suggest that, in order to keep pace with changes now being undertaken by the operational Navy and Marine Corps, the METOC community needs to reexamine many of its core approaches in a more systematic manner. **The de facto business model currently employed by the METOC community for providing enhanced warfighting capabilities should be examined and modified in light of e-commerce principles (e.g., peer-to-peer tasking and Web-enabled customer service) and network-centric warfare concepts of operations.** This review should be undertaken at four levels: customer interactions; data collection, data fusion, and information management; sensor networks; and network-centric operations.

To move forward and remain state of the art in environmental forecasting and prediction, resources need to be put into the development of both observational capability and models of processes that operate over small areas or over short time periods. There is a critical need to improve feedback in data and prediction flow between METOC and the customer. Failure to do so will compromise the ability of METOC to move forward and provide reliable predictions. The METOC community must build on existing relationships to strengthen its ties to operational U.S. Naval Forces and to the academic community, especially in the area of data assimilative forecast models. Furthermore, steps should be taken to familiarize warfighters, METOC and ONR personnel, and the academic community with the challenges faced by, and the strengths of, each of these communities. **The Oceanographer of the Navy, CNMOC, and the CNR should take steps to formalize efforts to transfer knowledge and experience among their respective commands and the academic community.**

---

[3]Use of the term "business model" may initially seem inappropriate to some readers in a report about supporting warfighters. This report, like other reports discussing business models, uses the term "business model" to describe the mechanisms and underlying philosophies that characterize a serious and organized endeavor. Thus, its use here is not intended to equate conducting war with conducting commerce.

Although uncertainty in environmental predictions introduces costs in terms of increased risk and occurrence of failure or the costs of contingency plans or suboptimal tactics, the benefit of better information must always be weighed against the cost of developing it. The goal of the METOC community is the reduction of environmental uncertainty in mission planning and operations. The optimum investment strategy is that which reduces environmental uncertainty to the level necessary for a desired probability of mission success and no more. Measures of the cost of uncertainty are not linear but are strongly concentrated on critical thresholds and are weighted differently for different variables and also by the value of the mission. **The Office of the Oceanographer of the Navy should promote reduction in the cost of uncertainty as a measure of value, so that a sensible strategy for research and development investments can be developed.**

For the research programs at ONR, the investment is not in augmented data collection but in the development of fundamentally better approaches. In contrast to the more applied METOC problem, there is no simple transfer function to equate the cost of research with expected return in reduced uncertainty. Thus, business methods for determination of the optimum rate and direction of investment cannot be readily employed. Instead, the cost of uncertainty provides an objective measure of research needs among many processes of disciplines that can drive the directions of research. Management wisdom must drive the decisions of funding rates and potential payoffs. **ONR should continue and expand efforts to understand and quantify both uncertainty and the cost of uncertainty in military operations. Furthermore, research priorities should incorporate an understanding of the relative impact uncertainty has on various naval operations, so that research priorities map to areas where the cost of uncertainty is the greatest.**

At present, the key to effective communication of environmental information to the naval commander lies with the attached METOC officer. These highly trained professionals play a complex role as synthesizer, filter, and interface. Increased use of advanced weapons systems (whose efficiency and effectiveness can be influenced by environmental conditions) and small naval units, such as Navy SEAL teams, that operate far from the fleet is changing the role of METOC. The growing need for integrated and organic environmental information systems to support weapons systems and small naval units will further stress a system that relies on interpersonal relations to maximize success. Furthermore, in order to more fully capture the benefits of improved measures of environmental uncertainty and the cost of that uncertainty, operational commanders need to more fully understand the accuracy of environmental information provided to them. The current naval concept of operations for understanding and assessing environmental uncertainty is contained primarily in the collective experience of METOC personnel, enhanced by informal and nondoctrinal infrastructures; thus, the quality and utility of environmental information will remain uneven and ephemeral, paced by the posting cycle of personnel. **The Office of the Oceanographer should work with U.S. Naval Forces operational commanders to introduce and explain the concept of environmental uncertainty and its value (including development of a common nomenclature for expressing uncertainty). U.S. Naval Forces in general, and the METOC community in particular, should take advantage of the concept of environmental uncertainty in more formal and recognized ways.**

1

Introduction and Overview

> This chapter introduces:
>
> - the history and current makeup of the naval meteorological and oceanographic (METOC) organization responsible for providing relevant environmental information to the U.S. Naval Forces (i.e., the U.S. Navy and U.S. Marine Corps),
> - the role of science and technology programs of the Office of Naval Research (ONR),
> - the major mission areas currently undertaken by U.S. Naval Forces and how environmental factors play a role, and
> - the goals of the study and the structure of this report.

June 2004, Indonesia: A ship suspected of carrying contraband arms has been tracked and is heading for the harbor of a hostile nation. A SEAL team is to be dispatched by high-speed boat to intercept, board, and take control of the vessel. The team leader checks with the METOC officer and ascertains that the wave heights are averaging 3 feet from the southwest. He estimates the intercept can be made a few miles before the ship enters hostile waters if the boat averages 45 knots. The mission is launched at 2300 hours. The sea has shifted to the northwest, and the wave height is now 6 feet. As the boat picks up speed, the violent shocks of the craft engaging the 6-foot seas directly on the bow become unbearable to the SEALs and the boat crew. The boat must slow to 30 knots and the intercept is missed.

May 2004, Gulf of Oman: An aircraft nuclear carrier (CVN) and three cruisers choose to anchor for the evening in the Gulf of Oman in the lee of Masirah Island. The anchoring process goes well, and the ships settle in for the night. Four hours after sunset, all of the ships start to experience clogging of their seawater cooling systems. Machinery starts overheating and air conditioning fails. Major electronic systems overheat and trip off line. A perceived small problem now threatens to disable the entire force. The engineers open the cooling pumps and find them clogged with what appear to be jellyfish. The staff METOC officer is called, and he recalls that this area is known for its cool water upwelling and an exceptionally high concentration of marine life. Further research reveals that in the evening layers of marine life are known to rise to shallow depths if attracted by lights. The ships order all external lights secured and the situation improves. Divers subsequently verify that a heavy layer of marine life rises in the evening and, if attracted by surface lights, comes almost to the surface.

February 2005, Somalia: An important naval gunfire support mission is called for during the early morning hours to support U.S. Marines deployed inland as part of a multinational force committed to breaking up a concentration of hostile combatants with ties to an international terrorist organization. The weather forecast predicts clear skies. As the shooting commences just after sunrise the spotter who is based on shore reports losing sight of the target due to a heavy haze that smells like wood fires. The mission continues without accurate spotting. A small number of men from the elite Canadian Airborne Regiment take advantage of a weakness in the enemy's perimeter defense and move forward. Radio communications between the troops inland and the surface force become unreliable. By the time news of the Canadian advance and new position are relayed to U.S. Navy surface combatants offshore, the Canadians have taken significant casualties from friendly fire. The diplomatic complaints are strong and endanger future joint operations in the area. Subsequent METOC analysis suggests that an early morning temperature inversion trapped the wood fire smoke that is common in developing countries and the strong atmospheric gradient resulted in electromagnetic ducting, which in turn resulted in poor radio communications.

U.S. Naval Forces carry out operations in the face of a complex set of challenges. Coordinating the actions of a large number of personnel and platforms, even during peacetime, requires a fundamental understanding of the capabilities and limitations of the assets involved and the environmental conditions under which activities will be carried out. Actions during wartime are even more challenging, as the actions and intentions of opposing forces significantly complicate every decision. Although environmental factors clearly will not always make or break military operations, warfighters, weapons systems, and platforms already under stress can be significantly and adversely affected by unforeseen or unplanned for environmental conditions. In the fast-paced and dangerous world of military operations, where anything that can go wrong often does, failure to understand the significance of environmental factors can have tragic consequences.

As pointed out by the three simple scenarios above, a naval commander must deal with some degree of uncertainty in almost every decision he or she makes. To support naval warfighters, the policymaker is challenged to decide how to allocate resources in a manner that reduces uncertainties in those areas where the result will do the most good. Is it important to know a target's location down to a foot or the depth of a harbor channel down to a few inches? In these cases, money would be better spent in other areas, since pinpoint target location and precise depth information are not needed to ensure success. METOC information clearly falls into a category where one must both evaluate the nature of uncertainty and determine acceptable levels of uncertainty. The challenge to the commander in the field is to understand the nature of uncertainty with respect to the various types of environmental information provided in order to conduct a broad spectrum of naval warfare operations. Infinite expenditures could come closer to yielding perfect information, but the stochastic nature of many environmental processes means that perfect information is not achievable. In addition, because environmental information is only one (and often a minor) factor that commanders must weigh, the ability to effectively utilize vast amounts of environmental information is limited.

Simple logic, therefore, suggests that some degree of uncertainty must be acceptable, as the cost of further reducing uncertainty through additional information gathering is prohibitive or the value of the improved knowledge it leads to is inconsequential in military terms. Setting priorities, therefore, for the acquisition, management, and dissemination of environmental information requires identifying what must be known and what level of uncertainty is acceptable under any given set of circumstances. Understanding the nature of uncertainty with respect to environmental data can be viewed as a form of tactical decision aid in that knowledge of uncertainty related to forecasts, models, sensor nets, and so forth can influence the choice of missions, weapons platforms, and "go/no go" decisions.

Short of having perfect environmental information, the naval commander needs to know how uncertain the information is in terms of time and space. In the METOC world, decisions regarding acceptable levels of uncertainty are often made years in advance by deliberations at the budget table or during weapons system design. Any analysis of the effectiveness of METOC information will lead to examination of a broad range of wartime and peacetime requirements and the recognition that perfect environmental knowledge (complete understanding of the processes shaping the environment) could yield better forecasts that are only marginally more valuable in many tactical situations than the forecasts that are currently obtainable. The analysis in the subsequent chapters focuses on areas where efforts to reduce uncertainty will produce the largest payoff in the quest for military superiority.

EVOLUTION OF THE ROLE OF METOC IN MILITARY OPERATIONS: THE PAST AS PROLOGUE

Mariners have always been very interested in forecasting weather and sea conditions. Initially this interest sprang simply from the desire to survive in a dangerous environment. High winds and heavy sea conditions could capsize ships or drive them far off course to a point where dwindling water and rations could be life threatening. Any advance warning could mean the difference between life and death. However, unreliable forecasts were of limited value and undermined confidence in forecasting.

The question was and is: Where should investments be made to improve reliability? Initially there were not many choices. As ships became larger and more seaworthy, sailors ventured greater distances and made observations about prevailing weather conditions, tides, currents, and water depths. It was no longer enough to survive. Economic success often depended on the skillful use of wind, weather, and currents. Masters developed charts, detailed sailing directions, and new instruments to help them avoid danger and reach their destinations before their competitors. Thumb rules such a "red skies in the morning, sailors take warning" abounded, along with sea tales that led to pilot charts and proprietary information created by shipping companies (see Box 1-1).

Knowledge of weather and sea conditions had other payoffs. In battle it could spell the difference between victory and defeat. From the beginning, there was always a strong military involvement in improved understanding of atmospheric and oceanographic processes. U.S. Naval Forces are the services most interested in maritime METOC. Working with the rest of the Department of Defense (DOD), the National Science Foundation, the National Aeronautics and Space Administration, the Department of Commerce, academia, and industry, the U.S. Navy makes a significant investment in and contributions to scientific efforts to understand fundamental METOC processes.

National Weather Service Under the Army

The evolution of the National Weather Service provides an interesting picture of who was most interested in weather forecasting. The first nationally organized weather service was developed under the direction of the military at what eventually became Ft. Myer in northern Virginia. The first director of the service was Brevet Brigadier General Albert J. Myer. The service was established in 1878 at Ft. Whipple, in northern Virginia in the hills overlooking the Potomac River and Washington, D.C. Organized forecasting was originally relegated to local observations and analysis, but the advent of the telegraph made possible the

BOX 1-1
Developing Forecast Capabilities: Skill Born of Necessity

The magnetic compass allowed ships to maintain roughly the right headings, and positions could be estimated by dead reckoning. Knowing one's location gave some clues to the expected weather and sea conditions, and changes in barometric pressure heralded changes in weather conditions. Cloud types, ocean swells, and wind direction gave more clues, and 18th-century masters could be quite proficient in assimilating data to form a reasonably accurate picture of what lay ahead. On sailing ships, a few minutes' warning to shorten or reef sails could avert disaster. The advent of the barometer gave captains the ability to foresee weather changes hours in advance. Admiral Robert Fitz-Roy of the Royal Navy systematically installed barometers on Royal Navy ships in 1854 and started the first heavy-weather warning system. The introduction of accurate clocks allowed the determination of longitude. This permitted more precise positioning and accurate charting, the avoidance of groundings, and a better correlation of historical weather and sea observations.

There was a wealth of knowledge in ships' logs that were compiled over the 18th and 19th centuries. It took a naval officer, Lieutenant Matthew Fontaine Maury, to collect, analyze, and publish these data in the form of pilot charts that provided seasonal data for the best shipping routes and military operations at sea (Nelson, 1990). This established the U.S. Navy and the Oceanographer of the Navy as the custodians of these data. Maury's effort to secure large amounts of data provided for development of general climatologies for many parts of the global ocean environment, a great step in reducing uncertainty.

The Office of the Oceanographer of the Navy has primary responsibility for meeting the environmental knowledge needs of the fleet. This is accomplished by a widespread network of METOC centers, survey vessels, and other remote and airborne assets, historical databases, computing facilities, and public/international environmental monitoring networks (e.g., World Meteorological Organization stations, U.S. Historical Climate). This environmental information is then distributed to forward-deployed battle groups as discrete products or packages of information that can be further analyzed by local METOC officers. The ONR works to develop understanding of relevant environmental processes, which can be applied through improved technological approaches to produce more accurate and useful environmental information for the fleet and Marine Corps.

Lieutenant Matthew Fontaine Maury is generally credited with the first U.S. efforts to collect, analyze, and publish nautical data in the form of pilot charts (Photo courtesy of the U.S. Navy).

transmission of observations and forecasts for a considerable distance. This led to the development of so-called synchronous reporting and synoptic weather charts and depictions of moving weather systems. Since the U.S. Army Signal Corps had the most robust communications network, it is no surprise that the original weather service developed under the Corps' direction. The obvious military advantages of accurately predicting the weather were not lost on Army tacticians. The Army was picked for this task because it was believed that military discipline would probably secure the greatest promptness, regularity, and accuracy in the required observations.

Weather Forecasts Important for Flight Safety

Knowing the weather at the origin, en route, and at one's destination was essential for safe flight. Before the days of instrument flying, aviation safety also demanded knowledge of conditions aloft, which led to the development of radiosonde balloons and the means to measure and forecast conditions at altitude.

By 1940 it was very clear to the Roosevelt administration that the largest and most demanding user of weather information was the aviation industry, which was rapidly surpassing railroads as the primary means of personal transportation and also becoming a significant freight carrier. For this reason the National Weather Service was transferred to the Department of Commerce, where it remains today.

Shift from Landline to Wireless Technology and Ship Safety

Landline telegraph and eventually wireless service were driving factors in the improvement of reporting and forecasting in general and specifically in dealing with weather conditions at sea. Wireless technology made it possible for ships to receive and send weather information to and from shore as well as to other ships. In 1902 shore stations first broadcast weather information to ships, and in 1905 ships were able to send their own data.

The importance of these developments cannot be overstated. Real-time observations from over 70 percent of the planet, which had previously been inaccessible, were now possible. The new system would soon grow to the point where scientists would reach the conclusion that weather is directly driven by sea surface temperature and thus climate is strongly influenced by thermal gradients and undersea currents.

Aviation-Capable Ships and Flight Safety

The operation of aircraft at sea demands accurate meteorological information from the surface to 50,000 feet out to hundreds of miles from aviation-capable ships. Reliable one- to six-hour weather forecasts are also critical for safe landing conditions. Today, aviation safety and efficiency are the dominant requirements for both the U.S. Navy and Marine Corps METOC communities.[1] Every large aviation-capable ship, shore facility, and fleet staff has its own METOC

---

[1]The Commander of the Naval Meteorology and Oceanography Command (CNMOC) heads the METOC effort within the U.S. Navy. While it is his responsibility to ensure that personnel within his claimancy serve the needs of U.S. Naval Forces, personnel in the Marine Corps METOC community are not under his command. Thus, when this report uses the term "U.S. Naval Forces" it is referring to operational forces of both the U.S. Navy and the Marine Corps (the end customers of the METOC enterprise). Conversely, when specifically referring to one or the other of the METOC communities, the terms "Navy METOC" or "Marine Corps METOC" are used as appropriate. This report is largely directed at examining and improving efforts by Navy METOC to support U.S. Naval Forces.

Rough seas can be a hazard even to today's large surface vessels such as this Burke class destroyer, the USS Mitchener (DDG 57), which displaces more than 8,000 tons (Photo courtesy of the U.S. Navy).

facility capable of providing detailed forecasts of any number of atmospheric conditions of interest to aviators.

Evolution from Safety to Warfighting Tool

It was soon recognized that a superior knowledge of water conditions could yield an advantage in battle. The requirement has thus evolved from safety to a serious warfighting tool. World War II saw a growing demand for knowledge of ocean conditions. The U.S. Navy, along with many other groups, began to understand that ocean conditions drive atmospheric conditions and that underwater characteristics determine acoustic properties vital to antisubmarine warfare.

Tactical cruise missiles with ranges of over 1,000 miles place even greater demands on METOC personnel, who must forecast en route and target weather conditions hundreds of miles away and up to 90 minutes in advance. The most complex problem of all is the land-sea boundary, where amphibious operations are conducted. Dynamic shore, beach, and undersea conditions generate complex environments that are very difficult to measure, model, or forecast. Any number of parameters, if incorrectly predicted, could threaten the mission. Many other factors need only be known to a general approximation. Thus, the key is not to predict all parameters with great accuracy but to predict the most significant ones with adequate accuracy. Discerning which parameters will be key is as difficult as developing the forecast.

Classic historical examples where METOC information was critical are the air war in Europe and the go/no go decision at Normandy. The long-range B-29 raids over Japan were aided by the discovery and prediction of high-altitude jet streams, and the final decisions on the atom bomb targets were driven by weather considerations.

LESSONS LEARNED

The short history given above brings us to the present, where the U.S. Navy maintains a METOC corps of 400 officers and over 1,300 enlisted personnel, stationed throughout the world. The U.S. Marine Corps maintains an additional METOC corps of about 450 personnel.[2] Their mission continues to be to measure, communicate, and forecast, with the highest certainty, weather and sea conditions. The remainder of this report examines, in part, how available resources are deployed in an efficient manner that ensures safety and military superiority at sea.

Clearly, the use of environmental information by U.S. Naval Forces has evolved dramatically. The coevolution of naval tactics and weapons systems with environmental observation and prediction capabilities continues in the 21st century. Combined with the dynamic nature of warfare operations, the naval battlespace is now recognized as a complex system where environmental conditions vary continuously on many temporal and spatial scales. Requirements for conducting modern military operations in the "4-D cube" (i.e., 3-dimensional space plus time) include the need for development of capabilities to observe and predict the global environment on spatial and temporal scales appropriate to overlapping warfighting operations (or mission areas). A continuing theme of environmental information needs is the corequirement to reduce or at least better understand uncertainty related to environmental data and forecasts.

MISSION DOMAINS IN MODERN U.S. NAVAL DOCTRINE

Modern U.S. Navy doctrine groups naval missions into five domains (Department of the Navy, 1997a, 2000). Four of these domains—sea dominance (e.g., mine warfare, antisubmarine warfare, surface warfare), air dominance (e.g., air

---

[2]Although U.S. Marine Corps personnel work closely with U.S. Navy METOC personnel in many instances, the Marine Corps' reliance on environmental information provided by the U.S. Navy varies.

B-29s of the U.S. Army Air Corps flew countless daylight missions over Japan. Accurate forecasts of the position of the jet stream played an important role in mission planning and execution (U.S. Army Air Corps photo).

defense, antiship missile defense, suppression of enemy air defenses), deterrence (e.g., deterrence of both conventional weapons and weapons of mass destruction), and power projection (e.g., strike warfare, naval special warfare, amphibious warfare)—represent traditional warfighter operations. The fifth area—sensors and information superiority (e.g., intelligence, surveillance, reconnaissance, naval METOC)—provides essential information needed to support the other four areas (Department of the Navy, 2002).

Tactical Oceanography

The value of oceanographic information for planning and executing naval operations has been recognized by the U.S. Navy for decades (National Research Council, 1997). Consequently, the ONR has been a primary source of funds for oceanographic research for many years. In an effort to improve the academic ocean science community's understanding of the operational demands placed on naval units, the Ocean Studies Board, through the support of the ONR and the

Office of the Oceanographer of the Navy, convened six symposia on tactical oceanography (National Research Council, 1991, 1992, 1994, 1996b, 1997, 1998, 2000).

These symposia, which focused on the role of environmental information for a variety of specific naval missions (e.g., amphibious warfare, antisubmarine warfare, strike warfare and ship self-defense, naval special warfare, mine warfare) constituted a valuable mechanism to facilitate more efficient use of naval research funds and to help academic scientists identify areas of research of high value to U.S. Naval Forces. A recurring theme in each symposium was the need for warfighters to receive timely and pertinent environmental information, specific to the needs of each mission. Real-time decisionmaking is crucial and the need for adequate and accurate environmental data on small scales is paramount for minimizing uncertainty and reducing risk.

U.S. Naval Forces operate in varied, dynamic, and often extreme environments; thus, accurate and detailed information on these areas is important for mission planning and operations. Personnel and platforms involved in naval surface warfare, aviation strike warfare, special forces warfare, amphibious operations, submarine and antisubmarine warfare, and counter-mine warfare operations have certain minimum environmental thresholds. When environmental conditions fall below these thresholds, performance is degraded. If these minimum thresholds are met, the success of the mission is partially the result of the ability to account for other higher-level and less predictable environmental variables; each area of operation has related but also other unique environmental criteria that are important. For example, the capacity to accurately predict environmental conditions at the launch platform, en route to the target, conditions at the target, egress from the target, and finally expected conditions at the return platform (especially at sea) figure significantly into the decision to proceed with an operation. Strike capabilities, antisubmarine warfare, ship defense, and special warfare operations in the littoral zone (where many future conflicts will most likely occur) are relying more and more on METOC's ability to describe and predict environmental conditions.

Over the past decade reports form this symposia series have highlighted the various environmental information needs for discrete mission roles, described certain knowledge deficits, and enumerated where future research should be directed. In general, environmental information is needed for the following roles: ship defense (including over-the-horizon defense, which thus expands the area for which environmental information is needed); target acquisition (e.g., infrared, laser, and electrooptical targeting); littoral penetration by special forces (e.g., swimmer and swimmer delivery vehicles to the surf zone, beach trafficability, and even inland road conditions); antisubmarine warfare (e.g., it is very difficult to detect and accurately target submarines in shallow nearshore environments due to such factors as acoustic backscatter, turbidity, and bottom characteristics that obscure active and passive detection); and counter-mine warfare (e.g., nearshore sediment transport as it interferes with mine detection), not to mention weapon delivery platforms, such as strike aircraft, surface ships, and submarines (e.g., environmental conditions influence how and where they operate as well as their stealth capabilities).

In 1993 the National Research Council released *Coastal Oceanography and Littoral Warfare*, which summarized lessons learned during a symposium on littoral warfare and highlighted a number of factors that affect naval operations (including mine warfare and amphibious warfare) in this zone. Table 2 of *Coastal Oceanography and Littoral Warfare* breaks these factors down into the following categories: atmosphere, biologic, oceanographic, bathymetric and topographic, acoustic, geophysical/magnetic, and anthropogenic. Environmental factors of importance that are found throughout this series of reports include:

Atmosphere
- Weather (clouds, fog, precipitation, wind speed and direction, air temperature)
- Ambient light, marine boundary layer properties (temperature, humidity, refractivity)

Biologic
- Ambient noise
- Optical scattering
- Bioluminescence

Oceanographic
- Tides
- Internal waves (currents: surface and subsurface)
- Water conductivity, temperature, depth, and salinity
- Sea state
- Wave height and direction
- Surf conditions
- Optical properties (vertical and horizontal)
- Turbidity

Bathymetric and Topographic
- Bottom and beach slope
- Beach and bottom composition

Acoustic
- Scattering
- Ambient noise

Geophysical/Magnetic
- Bottom roughness and type
- Sediment properties, bottom strength and stability
- Ambient magnetic and electrical background

Anthropogenic
- Pollution
- Noise

A follow-up report in 1997, *Oceanography and Naval Special Warfare: Opportunities and Challenges*, reinforced the previous report with a detailed list of areas in which METOC capabilities were inadequate and adequate but not optimal. The report also indicated that the current METOC capability to support naval special warfare is inadequate in the following areas: water temperature at depth, nearshore bathymetry, nearshore currents, lightning, internal waves, winds, precipitation (liquid), water clarity (turbidity), humidity (impact on communications as related to ducting or vulnerability), biofouling, beach trafficability, and bioluminescence. The following are adequate but not optimal: nearshore bathymetry, waves, tides, cloud ceiling, bottom composition, surf, offshore currents, visibility, and toxins and dangerous animals.

Additionally, conditions in shallow water environments are more subject to change than most other operational areas. Currents, tides, storm events, longshore transport of sediment, and other highly dynamic factors currently make planning these types of operations more of an art form than a science. High-resolution, temporal, and spatial data are needed if conditions are to be predicted with high confidence five to seven days in advance. Even then the chaotic nature of certain processes operating in these areas may limit both the precision and accuracy of five- to seven-day advance predictions for some parameters, regardless of how much data are collected. Nearshore bathymetry and beach condition change on small spatial scales and are difficult to measure, not to mention model.

In addition to the factors previously presented that will cut across many operational areas, the 1996 report *Proceedings of the Symposium on Tactical Meteorology and Oceanography* also identified three main areas for which METOC data are needed to support strike warfare. These are 100 km from shore, 100 km inland, and a 100-km radius around a ship. Temperature and humidity levels need to be better resolved vertically and horizontally, with higher resolutions near ground/sea level and progressively lower resolutions with increases in altitude. Data accuracy needs are also projected for:

Air temperature	(0.25°C)
Sea surface temperature	(0.50°C)
Relative humidity	(2 percent)
Wind vector	(10 percent)
Wave height and period	(10 percent)

Technology has always been a "force multiplier" for U.S. Naval Forces. The increased ability to measure these various environmental parameters, whether in situ or remotely, will impact future mission planning and operations in all theaters in which naval forces operate.

These challenges span global application of on-scene data collection; remote sensing data collection; integration of data collected in real time with archived data; incorporation of collected and archived data into the naval METOC production system; assimilation of collected and archived data into modeling and analysis efforts; and better assessments and predictions of METOC effects on platforms, weapons, and sensors in at-sea and real-time operations.

The mission of U.S. Naval Forces is to influence events onshore by projecting power from the sea. Since the end of the Cold War, naval operations have become increasingly focused in waters of the continental shelf and along coastal areas. This region, referred to by the U.S. Navy as the littoral (National Research Council, 1996b) has thus become the focus of efforts to understand and predict ocean and atmospheric processes and their influence on the conduct of naval operations. A thorough understanding of the coastal water column (depths of less than 100 m), the nature of the coastline, seafloor variability and stability, sub-seafloor characteristics, and the concentration of biological growth on or near the seafloor can help ensure mission success. In addition to concerns related to littoral oceanography, there is much that needs to be learned regarding air-sea interactions in littoral regions, air-sea-land interactions, atmospheric variability, and marine boundary layer dynamics (see Appendix B for more discussion).

Use of Environmental Information to Support Naval Missions

To help ensure success in these highly complex regions, higher-resolution descriptions of current conditions (nowcasts) and future conditions (forecasts), as well as analysis of oceanic, atmospheric, coastal, and beach conditions for littoral areas around the world, are sometimes needed. Capabilities to derive and/or measure data in denied areas and provide timely analysis and exploitation of current and developing technologies and available platforms have become essential. Warfighters require integrated synopses (tailored to their specific mission requirements) of a broad range of environmental parameters (see Chapter 2 for more detail). Efforts to improve the timeliness and usefulness of these environmental products will require not only an increased understanding of the environmental processes themselves but also parallel improvements in sensors and sensing capabilities, data management techniques, modeling approaches, computer software and hardware design, and increased communications capacity.

ORIGIN AND SCOPE OF STUDY

The current METOC enterprise and its predecessor organizations have brought the U.S. Naval Forces high-quality environmental information that has served it well in peace and in war. However, as the DOD transforms its force structure to meet the challenges that now face the nation, METOC must also examine how it will support the future. New approaches will be needed to provide METOC customers with information more rapidly anywhere and at any time. This will require new ways to collect the necessary data, new ways to analyze those data to create and present information, and new ways to deliver or make available that information worldwide to advantaged and disadvantaged users alike.

The METOC Organization and Relevant Efforts at the Office of Naval Research

The organization of U.S. Navy METOC (see Figure 1-1) is complex and characterized by the interaction of a robust Navy METOC officer, civilian, and enlisted organization with the operational U.S. Naval Forces at several levels. Tracing the lines of authority and the flow of money is difficult, and it takes considerable time and experience to grasp how the process works. Even with this challenging organization, METOC serves the U.S. Naval Forces extremely well and is highly regarded in government, industry, and academic worlds.

The Oceanographer of the Navy is an unrestricted line officer who normally comes to the assignment after an extensive career as a warfighter, commanding officer, and Navy resource manager. Recent oceanographers have been antisubmarine warfare specialists, nuclear submariners, and computer scientists. The oceanographer is normally not an oceanographer or weather specialist but rather acts as a bridge from the operational world of U.S. Naval Forces to the world of naval METOC. He or she is based at the Naval Observatory in Washington, D.C. This officer is actually a member of the staff of the Chief of Naval Operations (CNO) with the code N096. The budget authority stems from the CNO and includes all personnel, research and development,[3] and operating and ownership costs of the METOC community. The budget is managed by the N096 staff and executed by Navy METOC. In relative terms the entire budget is quite small and equates to less than half the cost of a new destroyer, or about $430 million.

Conversely, the Chief of Naval Research reports to the Secretary of the Navy through the Assistant Secretary of the Navy for Research, Development, and Acquisition (Figure 1-2). Thus, ONR constitutes a separately funded and administered science and technology effort totaling nearly $2 billion annually in programs throughout academia, industry, and the Naval Research Laboratories[4] (NRL, located principally in California, the District of Columbia, and Missis-

---

[3]The term "RDT&E" means research, development, test, and evaluation. It is the entire budget line for all nonprocurement programs. R&D, or RDT&E, includes science and technology (S&T) development efforts. "S&T" refers to 6.1 (basic research), 6.2 (applied research), and 6.3 (advanced technology development). N096 administers 6.4 (demonstration and validation) and 6.5 (engineering and manufacturing development) and operations and maintenance money. In effect, ONR is appropriated out of the lower half of the RDT&E account, N096 R&D out of the upper half.

[4]ONR's FY03 appropriation is about $2 billion, roughly $0.4 billion, $0.8 billion, and $0.8 billion, respectively, in 6.1, 6.2, and 6.3 programs. About 10 percent of the 6.1 and 6.2 total goes to NRL as "base funds"; the rest is competitively spent in academia, research labs, industry, etc. All 6.1-6.3 Navy S&T programs are administered through ONR.

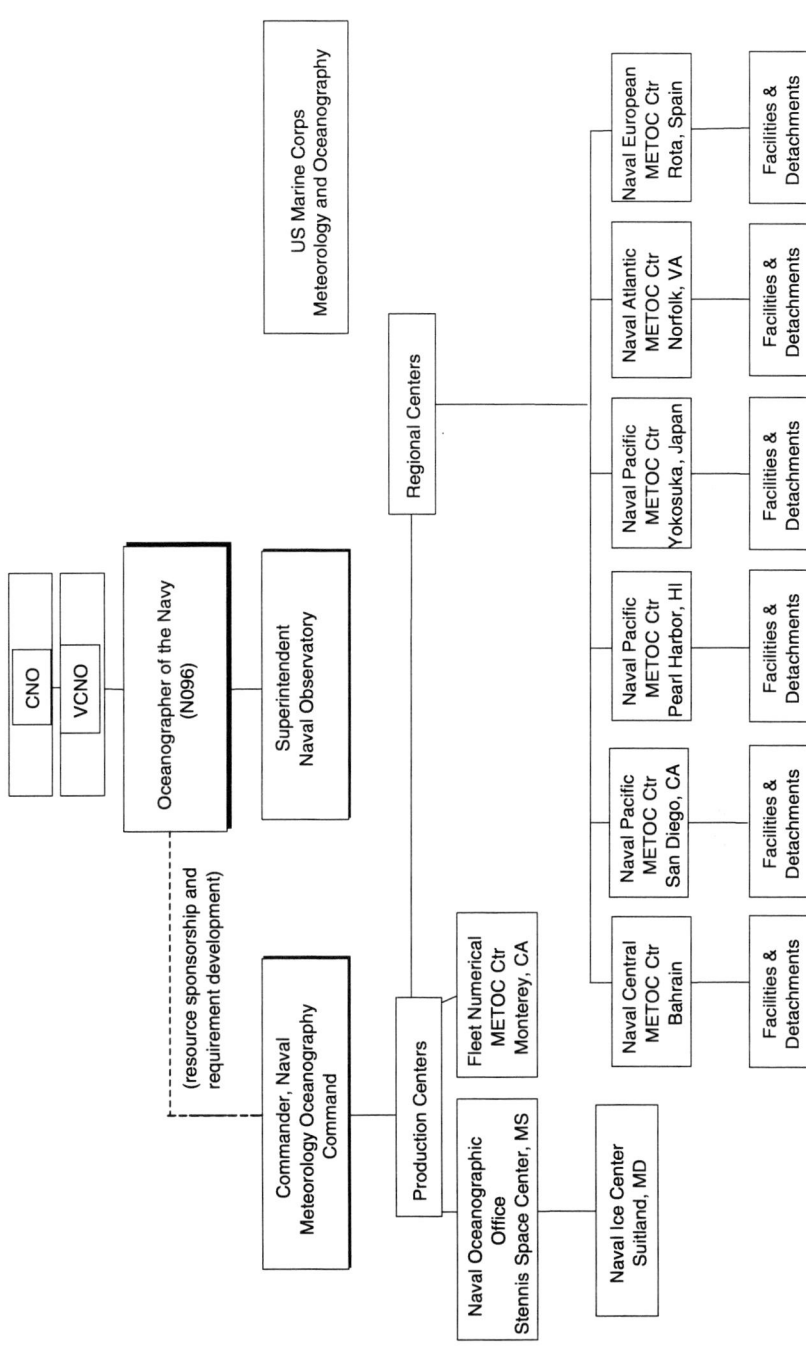

FIGURE 1-1 The Naval METOC Community (provided by the Office of the Oceanographer of the Navy).

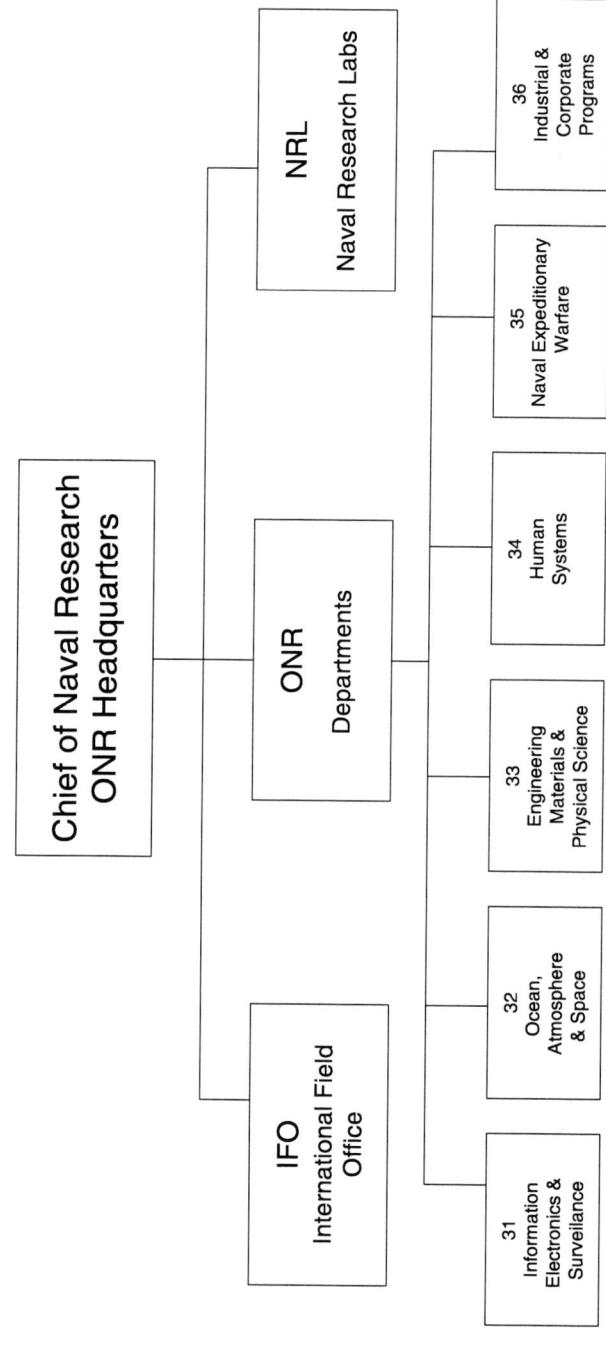

FIGURE 1-2 Organization chart for the Office of Naval Research (provided by the Office of Naval Research).

sippi). Thus, the U.S. Navy is one of the largest single supporters of both oceanographic and meteorological research.

Environmental science and technology programs at ONR are generically referred to as addressing aspects of battlespace environments as opposed to METOC (see Appendix C for discussion of relevant science and technology programs at ONR). Thus, throughout this text the use of METOC is restricted to programs and personnel actively involved in operational efforts (as opposed to science and technology efforts) to support the fleet and Marine Corps. While ONR is responsible for basic and applied research that may be of value to METOC efforts, ONR has no specific or formal relationship to the Oceanographer of the Navy, CNMOC, or the rest of METOC. For example, the Office of the Oceanographer's research and development dollars are directed at systems acquisition managed through the Space and Naval Systems Command. Conversely, ONR directly supports some warfare commands, such as the Naval Special Warfare Command. Shaping the nature of future efforts to address the need for environmental information by U.S. Naval Forces is thus a truly collaborative effort of several more or less autonomous groups within the Department of the Navy.

Navy METOC operations are the responsibility of the Commander of Navy Meteorology and Oceanography Command (CNMOC), who is physically located at the Stennis Space Center, near Bay St. Louis, Mississippi. This position is held by a career METOC officer holding the rank of rear admiral (lower half). Historically CNMOC serves for three years and then retires. Although only about 75 people make up the CNMOC staff, CNMOC oversees the entire METOC organization. The two key production centers are the Naval Oceanographic Office (NAVOCEANO), also at the Stennis Space Center, and the Fleet Numerical Meteorology and Oceanography Center (FNMOC) in Monterey, California. There are six fleet centers located in Norfolk, Virginia; Bahrain; Rota, Spain; San Diego, California; Pearl Harbor, Hawaii; and Yokosuka, Japan. Several other smaller activities such as the Naval Ice Center report to NAVOCEANO or directly to CNMOC (Department of the Navy, 1996, 1997b, 1999).

Personnel, procurement, and operations and maintenance money flows to CNMOC and out to the field activities. The Oceanographer of the Navy is the METOC program sponsor responsible for resources and requirements development, while CNMOC is the Navy METOC claimant and therefore the community overseer. The Oceanographer of the Navy is therefore a primary interface for CNMOC and the operational Navy. CNMOC's clamaincy is responsible for providing environmental support to the U.S. Naval Forces, but the Oceanographer of the Navy and CNMOC share responsibility for ensuring that the suite of products meets the needs of U.S. Naval Forces. U.S. Navy METOC, therefore, includes the worldwide survey fleet at this writing, consisting of eight ships that are operated by NAVOCEANO. This command has about 950 personnel and performs numerous functions in support of the operating forces and other components of METOC. These activities include maintenance of a world oceanographic data-

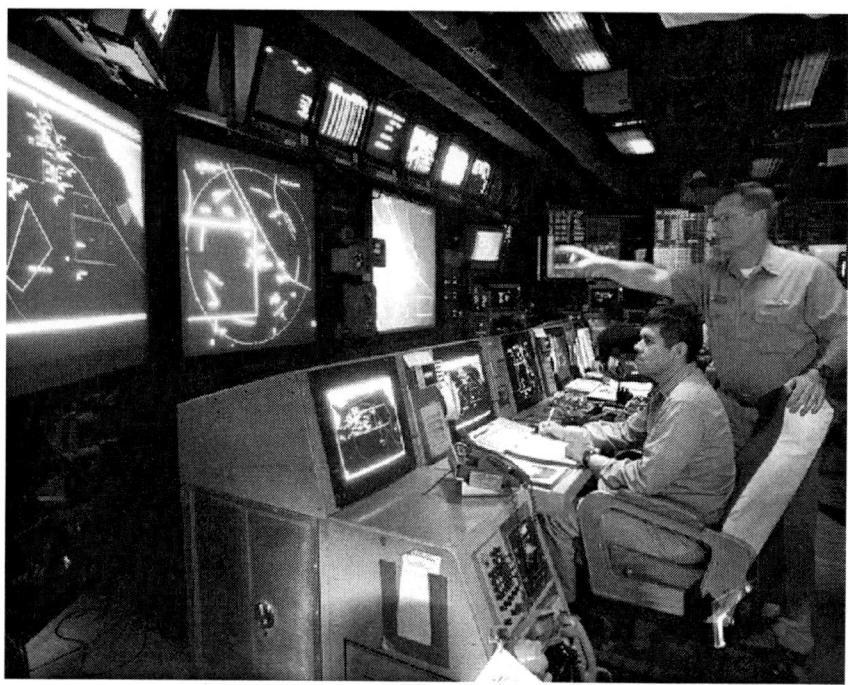

The U.S. Navy METOC enterprise produces environmental information to support U.S. Naval Forces. Modern combat information centers (such as the one depicted here) allow multisensors and information sources in establishing a common tactical information picture (Photo courtesy of the U.S. Navy).

base and extensive oceanographic models. The Warfighter Support Center at NAVOCEANO provides real-time and near real-time information to operating forces in the form of tailored reports and tactical decision aids.

The FNMOC has a staff of about 270 personnel and operates several supercomputer models of the atmosphere and oceans. It also functions as a communications hub that prepares and transmits products around the globe on a continuous basis.

The FNMOC's focus is on regional collection and forecasting and on providing direct inputs to the operational commander through local communications facilities. Some centers, such as the one in Hawaii, have special joint responsibilities. This center works with the U.S. Air Force to operate a Joint Typhoon Warning Center for the entire Pacific.

METOC personnel are also stationed on ships and stations throughout the world under various type commanders who operate aircraft, ships, and submarines. Their equipment and expendable resources are typically funded by these type commanders. The largest support is from the aviation community.

The METOC budget is reviewed every year, and a five-year plan is presented by N096 to the CNO. After priorities are set by N096, research and development money is provided to the Space and Naval Warfare Command in San Diego, where R&D projects are engineered to become operational systems. In an effort to ensure that fielded systems are truly of value to operators, the fielding of new systems must be funded with procurement and operational funds of the respective warfare communities. Thus, new METOC shipboard sensing systems have to compete with weapons, radar, and communications equipment. Because procurement and operational funds are hard to come by, this competitive process often delays modern METOC equipment installation. Achieving the support of the various warfare commands for new METOC capabilities is thus imperative.

Establishing priorities is further complicated by the competition for high-speed communications connections essential to the delivery of METOC products. The warfighting communities also fund this connectivity, and METOC officers must compete to obtain a significant portion of the communications "pipe." Fielding modern METOC capabilities is thus a complex, and at times divisive, decisionmaking process involving a variety of stakeholders with competing priorities. Over the years the METOC community has done remarkably well in making this process work. This is a reflection of the great skill and energy of the METOC officer and civilian leadership and supporting personnel. There is nothing automatic about the process. Budget battles must be fought and won every year.

The role of the Oceanographer of the Navy, especially one coming from a warfighting community, can therefore be critically important at the budget table. Interestingly, this assignment has been filled with surface and submarine officers almost exclusively. Over the past 15 years, there have been five surface officers and two submariners.

The dynamic process of establishing priorities for addressing the needs of U.S. Naval Forces for environmental information works well, but the complex interdependencies of budget support are cumbersome and delay introduction of the newest systems. Unfortunately, this is not unusual for large bureaucracies and certainly not unique to the U.S. Navy structure. It is only because science and technology are evolving so fast that these difficulties are now clearly denying our forces the latest technologies. It could be a crucial issue in the future.

THE TASK

In recognition of the growing need to support fleet operations in a variety of mission areas, the Office of the Oceanographer of the Navy and ONR requested that the Ocean Studies Board, in cooperation with the Naval Studies Board, undertake a two-year study to examine current and proposed efforts by the U.S. Navy to provide pertinent and useful information, in a timely manner, to the U.S. Naval Forces as a whole and to individual warfighters (as appropriate). Emphasis was placed on identifying the key characteristics of a process for optimizing the acquisition, assimilation, and application of meteorological and oceanographic data, including model development, fusion of data and value-added products with model results, and dissemination of environmental information (see Box 1-2).

STUDY APPROACH AND REPORT ORGANIZATION

It is a daunting task to develop insights and make useful recommendations for improving a process as complex and rooted in practical experience as naval METOC. Although the U.S. Navy and Marine Corps were very forthcoming in providing information on existing processes and programs (see Appendix E for a list of information-gathering activities conducted by the committee), the sheer volume of material took several months to assimilate. Based on this material, the committee focused on developing a few central themes that could in turn be used to refocus existing efforts or develop new programs. With that in mind, the present report is structured to reflect the development of these themes. Chapter 2 discusses the current METOC system and explores the value of the environmental information it produces. Chapter 3 provides a broad assessment of the current

---

BOX 1-2
Statement of Task

This committee will analyze the end-to-end environmental information system currently used by the U.S. Navy as well as that envisioned for the future to recommend possible approaches for improving both the individual components and the system as a whole. Emphasis will be on developing a framework process that can be adapted by the U.S. Navy to prioritize data collection and management, model development, fusion of data and value-added products with model results, and dissemination of environmental information to support individual missions and suites of naval missions. The committee will also identify segments of the process that would benefit from targeted research (e.g., specific ocean processes or general area of uncertainty). Finally, based on its analysis and the recommendations described above, the committee will prioritize the proposed improvements by identifying which actions are the most needed and achievable, which are most likely to make a needed impact on timeliness of analysis, and which are most readily exploitable given planned and available data collection opportunities.

state of METOC information and its sources and limitations. Analysis of this information will help the reader develop a sense of the existing METOC process and the key components of that process that the committee thinks could be modified to improve efficiency. Chapter 4 lays the groundwork for implementing changes to key components of the METOC system by focusing on uncertainty, the cost of uncertainty, and approaches to reduce both. Chapter 5 discusses the ramifications for expanded consideration and communication of uncertainty by implementing key network-centric concepts. Finally, Chapter 6 summarizes and categorizes the Committee's findings and recommendations in a manner that reflects the degree to which each is readily exploitable given planned and ongoing activities related to METOC or battlespace awareness.[5]

It was impractical for the committee to be exposed to all the programs and systems currently being developed to support U.S. Naval Forces. Thus, it is possible that many of the concepts discussed in this report are already being suggested, considered, or even implemented in some way or another. If the discussion in this report provides an impetus for deliberations, or if new concepts are evaluated and subsequently discarded in the face of more lengthy examination, the committee will believe its efforts have been worthwhile.

---

[5]Committee members found that the factors upon which they were tasked to prioritize their recommendations (described in the last sentence of the Statement of Task) were somewhat mutually exclusive. Thus, exploitability was chosen as the most appropriate organizing principle.

2

The Value of Environmental Information

> This chapter discusses:
>
> • why environmental information is of value to U.S. Naval Forces (the U.S. Navy and Marine Corps),
> • the primary components of the naval meteorological and oceanographic (METOC) enterprise, and
> • the underlying philosophies and approaches used by the naval METOC enterprise to serve U.S. Naval Forces
> • and provides the foundation for the key approaches used to evaluate the METOC enterprise in the remainder of the report.

Joint Publication 1-02 defines "battlespace" as the environment, factors, and conditions that must be understood to successfully apply combat power, protect the force, or complete the mission (Department of Defense, 1994). This includes the air, land, sea, space, enemy and friendly forces, facilities, weather, terrain, electromagnetic spectrum, and information environment within the operational areas and areas of interest. Since World War II, the operational area of U.S. Naval Forces has expanded globally to include all of the earth's marine environments (with increasing emphasis on the littoral region (i.e., the zone from 200 nautical miles offshore to the amphibious objective area) and the atmosphere. A thorough understanding of all aspects of the environment is critical to the mission success of U.S. Naval Forces.

The Chief of Naval Operations has assigned responsibility for meeting environmental information requirements to the Office of the Oceanographer of the Navy (N096). Thus, the mission of the N096 organization is to provide global METOC information, Precise Time and Astrometry (PTA), and Global Geospatial Information and Services (GGI&S) critical to operations conducted by U.S Naval Forces and, as needed, any joint or coalition actions that may be involved in (see Box 2-1).

BOX 2-1
The METOC Operational Concept

In May 2001 the Oceanographer of the Navy articulated a number of objectives that are the basis for the METOC Operational Concept:

- We will remain strategically engaged and become significantly more operationally and tactically aligned by providing decision makers with knowledge that is specific to the warfighting platform, system, weapon, etc., at the optimal time and place. METOC/PTA/GGI&S knowledge must exist inside the decision-making loop dictated by the speed of command.
- We will conduct a continuous assessment of the relevance and completeness of METOC/PTA/GGI&S, based on appropriate and accepted sets of metrics that are expressed in terms of operational or warfighting effectiveness (e.g., rounds on target).
- We will employ a doctrinal Rapid Environmental Assessment (REA) process, using both dynamic and static data to characterize the battlespace environment and reflect it in the 4D Cube as part of deliberate and contingency military operations.
- We will conduct near real time measurements, analysis, and "nowcast" of relevant METOC parameters throughout the battlespace, including denied areas. We will exploit all-source remote sensors and through-the-sensor methods to take advantage of existing warfare sensors and platforms. We will not develop or use stand-alone METOC sensor equipment, except when no alternative means exists to support the Operational Concept.
- We are the GGI&S experts of the Navy, and we will become information management experts. We will ensure the right METOC/PTA/GGI&S data are available to the right operator/network/system at the right time.
- We will recruit, train, equip, and retain personnel in a manner that best supports the Operational Concept. We will provide career profes-

continued

> **BOX 2-1 Continued**
>
> sional development that is synchronized with the Operational Concept and enables career progression opportunity.
> - We will determine the organizational infrastructure, resource infrastructure, and programmatic milestones that best support the Operational Concept.
>
> **Assumptions for the Operational Concept**
> - Network-Centric Operations/Warfare (NCO/W) will be predominant in the future.
> - Sufficient connectivity will exist to normally enable reliable exchange of data and information among various nodes on a robust network.
> - Remote sensing (including space, airborne, and undersea sensors/vehicles) will be the dominant, if not the only, source of information in denied access areas.
> - Through-The-Sensor Technology (TTST) will be available (to some extent) to support characterization of the battlespace environment.
> - Optimal manning in conjunction with NCO will drive a redistribution of manpower and expertise to realize robust reach-back capability and virtual presence.
> - Speed of command will be paramount, requiring more automation of processes.
> - The 4D Cube (geo-reference to WGS-84 and time-reference to UTC) will provide the information and knowledge basis for common operational pictures.
> - Enhancing "safety of operations" and providing "warfighting advantage" will remain critical mission requirements for Naval Oceanography.
> - Objective metrics based on customer impact will be used to evaluate the effectiveness of METOC/GI&S/PTA support.
>
> SOURCE: Department of the Navy (2002).

The diversity of naval warfare missions is illustrated by Table 2-1. Given the broad spectrum of modern naval operations, environmental information requirements to successfully conduct these missions are also remarkably diverse (see Table 2-2).

In meeting the demands of naval mission areas, the naval METOC community provides operational support resources to fleet commanders and their subordinate commands. The Oceanographer of the Navy is responsible for obtaining the resources needed to build and maintain the data, models, and other tools

TABLE 2-1 Naval Mission Areas Included in the General Requirements Relational Database

Mission Area	Acronym
Anti-Air Warfare	AAW
Amphibious Warfare	AMW
Anti-Surface Warfare/Over-the-Horizon Targeting	ASW/OTHT
Command/Control/Communications/Computers, Intelligence, Surveillance, and Reconnaissance	C4ISR
Logistics and Sealift/Joint Logistics Over the Shore	LOG/JLOTS
Mine Counter Measures/Mine Warfare	MCM/MIW
Operations Other Than War	OOTW
Naval Special Warfare	NSW
Strategic Deterrence and Weapons of Mass Destruction	STRAT/WMD
Strike Warfare	STRIKE
Undersea Warfare	USW
Wargames and Training Issues	WGT

SOURCE: Handfield and Clark (1999).

necessary to monitor and predict environmental conditions of importance to each of the mission areas and those operations contained within the mission areas. The main engines for environmental information flow, modeling, and forecasting are two large computational centers ashore: the Naval Oceanographic Office (NAVOCEANO) and the Fleet Numerical Meteorology and Oceanography Center (FNMOC).

DATA COLLECTION

Traditionally METOC has not possessed a robust in-house observational network to rely on to meet its operations requirements, tending rather to rely on synoptic sensor data from a variety of external sources such as the World-Wide Weather Watch, the National Oceanic and Atmospheric Administration's (NOAA) buoy data and ships of opportunity, NOAA and Department of Defense (DOD) satellite data, and historical archives maintained by NOAA's Climate Center and NAVOCEANO archive data from TAGS ship collections. These synoptic data are assimilated at the two computational centers and used as input to large-scale numerical models, which are run on a routine daily schedule.

NAVOCEANO operates a fleet of oceanographic survey ships to collect information from the seafloor, ocean surface, and water column. The TAGS50 and TAGS60 class ships provide multipurpose oceanographic capabilities in coastal and deep-ocean areas, including physical, chemical, and biological oceanography; multidisciplinary environmental investigations; ocean engineering and

TABLE 2-2 Incidence of Environmental Parameters Within the 12 Naval Mission Areas (from Table 2-1 Above) as Reported in the General Requirements Database (GRDB)[a]

Environmental Parameter	No.	Environmental Parameter	No.
Blowing sand	12	Refraction (optical and acoustic)	9
Ceiling layers	12	Topography	9
Cloud cover, type, etc.	12	Twilight time	9
Dew point	12	Bioluminescence	8
Ducting	12	Breaker interval	8
Aerosols, haze, smoke	12	Salinity	8
Humidity	12	Temperature (water column profile)	8
Icing (aircraft)	12	Tides (phases, heights, times, currents)	8
Lightning	12	Water clarity (turbidity)	8
Moisture profile	12	Breaker direction	7
Precipitation (type, rate, total)	12	Breaker height	7
Sunrise/sunset	12	Ice edge	7
Temperature (air, sea, land)	12	Marine mammals	7
Temperature (vertical profile)	12	Moon angle	7
Wave height, period, direction, etc.	12	Moon illumination	7
Surface wind (direction and speed)	12	Moon phase	7
Barometric pressure	11	Noise (precipitation)	7
Fog	11	Reverberation effects	7
Icebergs	11	Sound speed profile	7
Icing (sea surface)	11	Sun angle	7
Magnetic anomalies	11	Time interval synchronization	7
Moon rise/set	11	Transmission loss	7
Sea ice	11	Astronomical time	7
Sea spray	11	Beach slope	6
Water depth	11	Bottom gradient	6
Wind aloft (direction and speed)	11	Bottom loss	6
Wind shear (i.e., vertical wind profile)	11	Bottom roughness	6
Air turbulence	10	Breaker type	6
Currents (surface)	10	Clutter density	6
Inversion rate	10	Currents (bottom)	6
Ionospheric scintillation	10	Frequency stability	6
Precise time	10	Noise (shipping)	6
Solar flares	10	Noise (waves)	6
Solar flux	10	Shipping traffic	6
Temperature (horizontal variation)	10	Convergence zone	5
Anchorages	9	Noise (sea ice)	5
Seafloor composition	9	Beach trafficability	5
Reefs	9	Noise (biological)	4
Currents (water column)	9	Surf zone length	4
Precipitation (type)	9	Surf zone width	4

[a]No. column refers to number of naval mission areas in which parameter was listed as having an influence on planning/execution of the mission.

Naval personnel ensure that a chemical explosive device has been properly "decontaminated" during an exercise by checking for vapors. The exercise, Tri-Crab 2001, was a multinational, joint service, explosive ordnance disposal (EOD) training exercise involving teams from the U.S. Navy, U.S. Air Force, Royal Australian Navy, Royal Australian Air Force, and the Royal Singapore Navy (Photo courtesy of the U.S. Navy).

marine acoustics; marine geology and geophysics; and bathymetric, gravimetric, and magnetometric surveying.

In situ METOC measurements and observations are also a contributor to warfighting support products. These can be collected by either METOC personnel or, more commonly, the asset commander. For example, a deployed submarine may collect in situ information on the water column that can be used to improve a local model output if it were fed back to the METOC centers. The OA[1] Division and the Mobile Environmental Teams are equipped to collect organic data. These METOC personnel collect local surface observations and upper-air soundings that can be transmitted back to FNMOC via the Automated Weather Network (AWN). In addition, the carriers have satellite full-resolution direct-readout capability, and the Mobile Environmental Teams have a reduced-resolution satellite readout capability. The OA officer afloat can also access environmental information from a wide variety of third-party providers such as foreign weather services, research and commercial satellite imagery, information from universities, and other research centers worldwide.

Remote Sensing

The U.S. Navy's remote sensing program focuses on tactically significant information in the littoral regions. Environmental sensors on existing satellites are being exploited to the greatest extent possible, and new capabilities, such as those offered by WindSat and hyperspectral imagers, are under development. Remote sensors on airplanes, remotely piloted vehicles, surface ships, and undersea systems appear to be underutilized, as are environmental data that can be extracted from sensors deployed for other-than-environmental missions.

Modeling

Model output, in the form of gridded data fields, is the primary product from the two Navy computational centers. These fields are distributed by a variety of means, one of the most significant being the DOD's AWN, operated by the U.S. Air Force. The gridded products are used by several DOD and civilian agencies, such as Defense Threat Reduction Agency for chemical dispersion modeling and NOAA as a backup hurricane forecast. Most importantly, they are used by the six METOC centers around the world to prepare ship routes and aviation forecasts.

Both the computational centers and the six regional centers can use the gridded output to run mesoscale models, such as the Coupled Ocean-Atmosphere Mesoscale Prediction System (COAMPS). COAMPS is a short-term (nowcast to 48-hour) mesoscale forecast tool for the littoral. A gridded mesoscale model out-

---

[1]The acronym "OA" refers to the military designation "Operations, Aerography" and dates back to a period in naval history when METOC personnel were rated as aerographers.

The U.S. Navy's Fire Scout Vertical Takeoff and Landing Tactical Unmanned Aerial Vehicle (VTUAV) launches into its flight test program at the Naval Air Forces Western Test Range Complex in California. Fire Scout has been designed to provide situational awareness and precision targeting support for U.S. Naval Forces, as a fully autonomous UAV requiring limited human intervention. Future operations are planned for the UAV during the summer of 2003 (Photo courtesy of the U.S. Navy).

put can be forwarded to the OA division on aviation-capable ships, but typically regional center forecast products are delivered to customers as visual aids or text messages.

Forecasting

Direct forecasting support to staffs, ships, and units afloat and onshore is provided by officers and enlisted METOC personnel assigned to these activities. Permanently embarked METOC personnel and deployable assets (Mobile Environmental Teams) provide on-scene support for forces afloat and those in-theater onshore. The U.S. Navy's permanently afloat METOC organic assets are its OA divisions, embarked onboard major aviation-capable combatants and command ships. The primary objectives are safety, optimum tactical support to warfare commanders, and tailored on-scene products and services.

Information flow for forecasting services is shown in Figure 2-1. The process begins with a standing order from the user community for a forecast. Data are assimilated, models initialized and run, and gridded outputs delivered to cus-

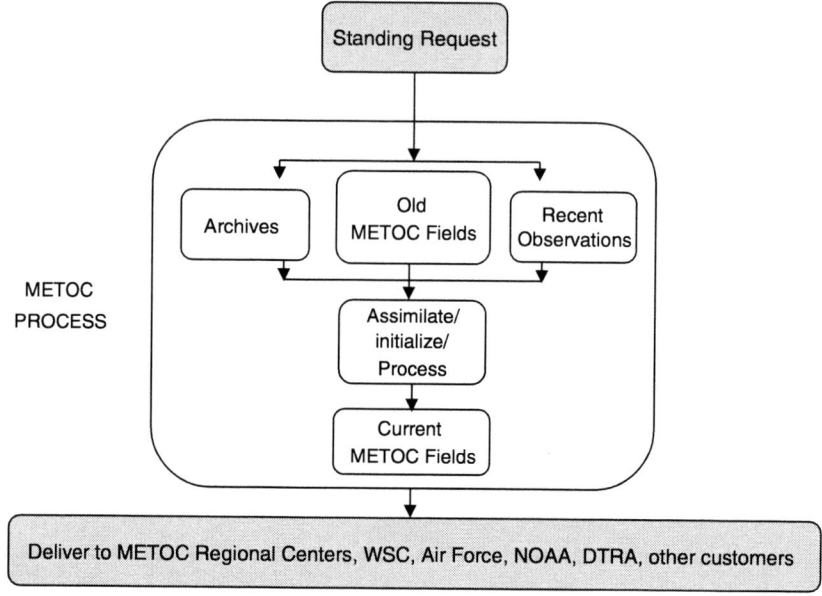

FIGURE 2-1 Diagram illustrating the typical one-way flow of synoptic and mesoscale gridded model output within the existing METOC system.

tomers. Typically, the forecasting process ends with METOC personnel providing a briefing or consultation with an asset commander. Forecasting services are essentially "stovepipe" services, meaning that the asset commander sees the output but has no intermediate data access or visibility into the process. The asset commander requests a product and receives it. Furthermore, there is presently no provision for customer interaction and feedback during the numerical process or to benchmark current performance of model output forecast products and tactical decision aids or other enhanced warfighting products.

Mission Objectives

The U.S. Navy METOC Strategic Plan implemented by N096 (The Department of the Navy, 1999) lists four mission objectives:

1. Optimized warfighting resources—Generate fiscal savings and increase military readiness through better forecasting of the global environment.
2. Advanced core competencies—Influence and exploit advances in science and technology to best characterize the natural environment.
3. Safe operating forces—Protect all assets.

4. Enhanced warfighting capabilities—Fully characterize the battlespace environment to the warfighter in terms that enable optimal employment of systems and platforms.

The first two mission objectives are internally focused, while the latter two have direct interface to operational commands and immediate impact on Navy warfighting missions (Table 2-1). As emphasized in the Statement of Task (see Chapter 1), this study's focus is to examine the METOC community's effectiveness in meeting the latter two externally focused missions in order to assist in optimizing strategies for meeting its internal mission objectives. Although the subject of much interest to individuals charged with monitoring fiscal resources, the objective to optimize warfighting resources differs so fundamentally from the other three objectives that tend to focus more closely on warfighter needs that the committee chose not to explore it in any detail.

Advanced Core Competencies

A primary focus of naval METOC functions is to provide timely environmental data and predictions to warfighters (Alberts et al., 1999). To achieve these objectives, the naval METOC system must be sufficiently robust and flexible to meet the information requirements of both day-to-day operations and specific demands of national security (driven by the specific objectives of a given operation) on a variety of spatial and temporal scales.

Research is conducted in four general areas: (1) physical understanding of environmental processes; (2) model development and prediction; (3) observations for prediction of parameters important to operational safety and military mission success, and (4) systems analysis, including the collection, fusion, and dissemination network. Crosscutting these areas are the three regions of the earth's environment (atmosphere, land, and oceans), together with their coupling and interactions.

Research supported by the Navy intended to directly benefit "safe operating forces" traditionally focuses on improving the range and accuracy of synoptic and mesoscale forecasts. Research for "enhanced warfighting capabilities" is more diverse than for forecasting services and can include sensor array options, physical understanding of the environment, and distributed systems architectures, as well as model improvements.

Safe Operating Forces

Since the Navy typically operates in a peacetime environment, "safe operating forces" tends to capture the largest share of the community's resources. The primary products meeting the safe operating forces mission objectives are derived from synoptic and mesoscale atmospheric and oceanographic model output.

Examples include Optimum Track Ship Routing (OTSR) and Optimum Aircraft Routing System (OPARS) forecasting services. However, as mentioned above, increases in efficiency due to improved observation, modeling, and forecasting are likely to result in only incremental improvements in efficiency in this area. Since naval capabilities in this area are sufficient and expected to remain so into the foreseeable future, increased investment in improved OTSR or OPARS is not recommended.

Enhanced Warfighting Capabilities

The product line for supporting the third N096 mission objective, enhanced warfighting capabilities, includes geospatial products, such as Special Tactical Oceanographic Information Charts (STOICS) and Special Annotated Imagery and model output from specialized tactical decision aids such as acoustic propagation prediction, ocean front and eddy thermal analysis of the ocean surface, or

A U.S. Navy SEAL provides cover for his teammates while advancing on a suspected location of al Qaeda and Taliban forces. Supporting small, forward-deployed U.S. Navy special operations forces conducting missions in Afghanistan in support of Operation Enduring Freedom presented unique challenges to the U.S. Navy's METOC community and Office of Naval Research (Photo courtesy of the U.S. Navy).

a nowcast of electromagnetic ducting conditions in the atmosphere's boundary layer distortions). Products designed to enhance warfighting capabilities tend to be localized and focused on the littoral regions, with finer spatial resolution than the forecasting services described above for safe operating forces.

Environmental support for the warfighter is typically initiated at the Warfighting Support Center co-located with the NAVOCEANO computing center at Stennis Space Center, in Mississippi, or through one of the regional centers. Weapons systems performance can be affected by environmental conditions; thus, the Oceanographer of the Navy also has military and civilian staff assigned to each of the major systems commands—Naval Air Systems Command, Naval Sea Systems Command, and SPAWAR.

The information flow for preparing enhanced warfighting products is shown in Figure 2-2. As with the METOC synoptic and mesoscale forecasting products described earlier, warfighting support products are delivered through a traditional stovepipe process. The warfighter has no continuing input to, or information from, the process. Although some products, like STOICS, are offered via Web services as a smorgasbord of information downloadable by the warfighter, use of network technology for METOC for full customer interaction and product distribution is limited.

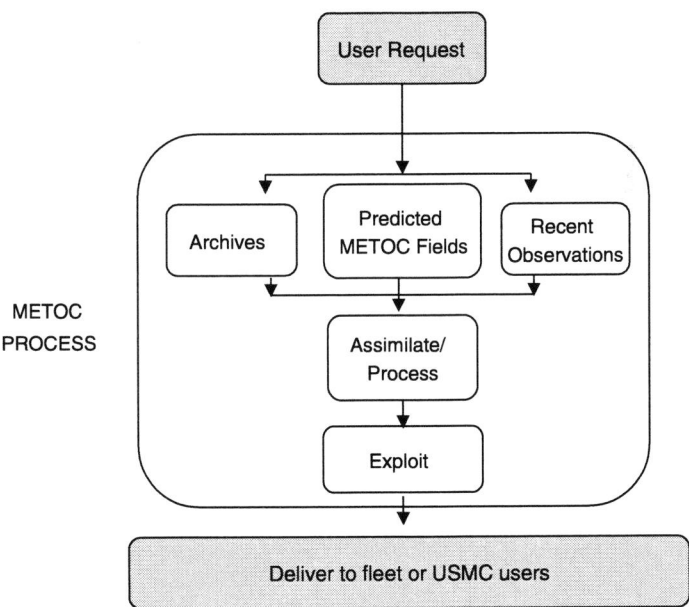

FIGURE 2-2 Diagram illustrating the typical one-way flow of information associated with Enhanced Warfighting Capability within the existing METOC system.

Furthermore, communications bandwidth is limited on most surface combatants other than carriers and is very small and often intermittent on submarines and in forward-deployed Marine units. There are no METOC personnel on submarines or small combatants; hence, the level of METOC training is different. Generally the flow is one way through the system from tasking to delivery. Opportunity for feedback is limited, and feedback mechanisms are not designed into the current system.

Warfare Areas

The principal naval warfare areas and the environmental parameters of importance to those warfare areas are tabulated in Table 2-3. Increasingly, there is a need for environmental information on finer spatial and temporal scales, including in some instances the requirement for near real-time environmental assessment in support of time-critical operations. However, as the spatial and temporal scales of environmental information are compressed, uncertainties regarding the accuracy, timeliness, coverage area, and assimilation of data expand. Further, many presently available METOC products (such as six-hour forecasts) are not sufficient for the information needs of rapidly changing battlespace conditions. Thus, uncertainty regarding the state of the environment remains a critical limitation to enhanced warfighting capabilities. There is an important and growing need in the METOC community to develop meaningful assessment and evaluation metrics to characterize uncertainty in environmental information and to determine the impact of that uncertainty on naval operations. At present, the general requirements for environmental information are developed through a complex and often cumbersome process. The role of warfighters and users in developing quantitative estimates for some critical thresholds, or other numerical expression of platform or personnel sensitivity, is limited. This is somewhat understandable, as much information gained through such exchange is allegorical in nature and difficult to quantify. However, as more robust methods for capturing warfighter experience are developed and as weapons systems with specific and clearly expressed environmental tolerances become more widespread, quantifying the needed accuracy of environmental information will become easier. At the same time, there will be a greater need to provide indoctrination and training of METOC personnel in the characterization and understanding of uncertainty such that corresponding analyses of risks associated with those uncertainties can be conveyed to warfighters and asset commanders.

STRATEGIES FOR TARGETING INVESTMENT IN RESEARCH AND DEVELOPMENT

The gauge of success in business is straightforward—creation of profits through the sale of popular products or services. Success is a bottomline figure—

TABLE 2-3 Relevance of Environmental Parameters to Naval Warfare Areas[a]

Environmental Parameter	AMW/ NSW	ASW/ USW	MIW/ MCM	STRAT/ WMD	STRIKE	MULTI
Aerosols, haze, smoke	X	X	X	X	X	X
Air turbulence	X	X	X	X	X	X
Anchorages	X	X	X	X		X
Astronomical time	X	X	X	X		X
Barometric pressure	X			X	X	X
Beach slope	X	X	X	X		X
Beach trafficability	X	X	X	X		X
Bioluminescence	X	X	X	X		X
Blowing sand	X	X	X	X	X	X
Bottom gradient	X	X	X	X		X
Bottom loss	X	X	X	X		X
Bottom roughness	X	X	X	X		X
Breaker direction	X	X	X	X		X
Breaker height	X	X	X	X		X
Breaker interval	X	X	X	X		X
Breaker type	X	X	X	X		X
Ceiling layers	X	X	X	X	X	X
Cloud cover, type, etc.	X	X	X	X	X	X
Clutter density	X	X	X	X		X
Convergence zone	X	X	X	X		X
Currents (bottom)	X	X	X	X		X
Currents (surface)	X	X	X	X		X
Currents (water column)	X	X	X	X		X
Dew point	X	X	X	X	X	X
Ducting	X	X	X	X	X	X
Fog	X	X	X	X	X	X
Frequency stability	X	X	X	X		X
Humidity	X	X	X	X	X	X
Ice edge	X	X	X	X		X
Icebergs		X	X			X
Icing (aircraft)	X	X	X	X	X	X
Icing (sea surface)	X					X
Inversion rate	X			X	X	X
Ionospheric scintillation	X				X	X
Lightning	X	X	X	X	X	X
Magnetic anomalies	X	X	X			X
Marine mammals	X	X	X	X		X
Moisture profile	X	X	X	X	X	X
Moon angle	X	X	X	X		X
Moon illumination	X	X	X	X		X
Moon phase	X	X	X	X		X
Moon rise/set	X				X	X
Noise (biological)	X	X	X	X		X
Noise (precipitation)	X	X	X	X		X
Noise (sea ice)	X	X	X	X		X

continued

TABLE 2-3 Continued

Environmental Parameter	AMW/ NSW	ASW/ USW	MIW/ MCM	STRAT/ WMD	STRIKE	MULTI
Noise (shipping)	X	X	X	X		X
Noise (waves)	X	X	X	X		X
Precipitation (type, rate, total)	X	X	X	X	X	X
Precise time	X	X		X	X	X
Reefs	X	X	X	X		X
Refraction (optical and acoustic)	X	X	X	X		X
Reverberation effects	X	X	X	X		X
Salinity	X	X	X	X		X
Sea ice		X	X			X
Sea spray	X	X	X		X	X
Seafloor composition	X	X	X	X		X
Shipping traffic	X	X	X	X		X
Solar flares	X	X			X	X
Solar flux	X	X			X	X
Sound speed profile	X	X	X	X		X
Sun angle	X	X	X	X		X
Sunrise/sunset	X	X	X	X	X	X
Surf zone length	X	X	X	X		X
Surf zone width	X	X	X	X		X
Surface wind (direction and speed)	X	X	X	X	X	X
Temperature (air, sea, land)	X	X	X	X	X	X
Temperature (horizontal variation)	X	X	X	X		X
Temperature (vertical profile)	X	X	X	X	X	X
Temperature (water column profile)	X	X	X	X		X
Tides (phases, heights, times, currents)	X	X	X	X		X
Time interval synchronization	X	X	X	X		X
Topography (land)	X	X	X	X		X
Transmission loss	X	X	X	X		X
Twilight time	X	X	X	X		X
Water clarity (turbidity)	X	X	X	X		X
Water depth	X	X	X			X
Wave height, period, direction, etc.	X	X	X	X	X	X
Wind aloft (direction and speed)	X	X	X	X	X	X
Wind shear (i.e., vertical wind profile)	X	X	X	X	X	X

X = important parameter.

[a]AMW = Amphibious Warfare; NSW = Naval Special Warfare; ASW = Antisubmarine Warfare; USW = Under Sea Warfare; MIW = Mine Warfare; MCM = Mine Counter Measures; STRAT = Strategic Defense; WMD = Weapons of Mass Destruction; STRIKE = Naval Strike Warfare; MULTI = Multi-Mission Scenarios. Shaded = important parameter; Not shaded = unimportant parameter.

SOURCE: Handfield and Clark (1999).

the net profit from operations. Thus, inherent in business is an objective value system that forms a clear basis for planning.

While most businesses invest in Research and Development, no major business would proceed with such an investment without a full business plan. Such a plan consists of three components. The first is a financial model for payback on investments through later profits. It represents the best guess of planners on the response of the future market to the developed products. The second, and equally important, aspect of planning is an analysis of risk. It is recognized that the payback model contains many assumptions whose failure could have serious consequences. Probabilities must be estimated and the financial impact of various scenarios computed. Finally, the risk posture of the company must be assessed and an acceptable level determined. The final investment strategy, then, is that which provides the optimum payback within the level of risk that is acceptable to the business.[2]

Application of Business Principles to Naval METOC Operations

Development of an objective basis for investment in METOC by the Department of the Navy requires application of the same business principles and logic (e.g., cost-benefit or risk posture analysis) to the METOC problem.[3] This, in turn, requires development of an appropriate value system for naval activities and the impact of METOC on those activities. It also requires a risk analysis to quantify the potential impact of chance on operations. Finally, an assessment must be made of the risk level that is acceptable in expected operations. With these elements quantified, objective decisions can be made.[4]

In considering the value of METOC, it could be concluded that improvements in METOC capability have value in two ways. A direct but small value comes from efficiency gains such as Optimum Track Ship Routing (OTSR), resulting from improved knowledge of the environment. This might yield significant savings each year. The second value comes from risk reduction. Even if only a very small fraction of operations were impacted through improved risk management, this value would dwarf that from optimized resource applications such as OTSR. The costs of lost aircraft or ships, of lost lives, or of failed missions with possible hostage consequences are enormous. Thus, the following

---

[2]Conceivably, a business may conclude that the risks are too poorly known and invest in further market or other research to reduce its uncertainty.

[3]For example, one could argue that recent trends toward emphasizing applied (6.2) research at ONR reflect a change in the risk tolerance of that organization. Basic (6.1) research is a longer-term investment that may or may not lead to significant Navy applications in the near term, whereas 6.2 research is more likely to yield near-term benefits.

[4]By corollary, no objective decisions can ever be made without these assessments. When past investment decisions have been made in the absence of a formal analysis, they have been based on an individual's judgment of the various values.

METOC briefings are a regular component of the premission preparation for U.S. Navy and Marine pilots. The intimacy and regularity of these briefings represent a unique interaction between information provider and warfighter. In most other naval mission areas, environmental information is provided periodically and by electronic means. In such situations, greater effort must be made to capture feedback from the warfighter on the timeliness and value of the information received (Photo courtesy of the U.S. Navy).

discussion focuses on the role of METOC in naval risk management, leaving resource maximization efforts as secondary and already well handled.

The application of business principles to military operations is not straightforward. How should a value be placed on freedom, peace, or open commerce? How well can risk from environmental factors be quantified in military operations? How much risk from environmental factors should operational commanders be willing to accept? Assessment of these matters can be very difficult. However, by attempting to bound the problem, a number of useful points arise.

Risk Within the Military

While risk is inherent in all military operations, there is a clear and, in recent years, high value placed on risk reduction in military operations. Mission failure not only results in the nonaccomplishment of an important task but also draws the attention of a public and press that are overly accustomed to winning. The loss of combatants and even the presence of collateral civilian casualties can have a compromising effect on the very foreign policy that military action is intended to help implement. Perhaps even worse is the situation of military or civilian hostages. The acceptable cost of hostage rescue is huge, as can be the acceptable compromise of policy for hostage rescue purposes. Even acknowledgment of environmental uncertainty in planning brings associated costs in the form of costly contingency plans or low-risk suboptimal mission plans.

Risk in military operations comes from two primary sources. The first is the unpredicted activities of the enemy (e.g., change in location, mounting a defense, counterattack). This is mitigated, usually successfully, through reconnaissance gathering. The second source comes from the unpredictability of the environment. It is this source of uncertainty that is addressed by the METOC enterprise and will also be the focus of the remaining discussion.

In a business model, risk assessment requires quantification of the probabilities of certain risks and their associated financial impacts. From the cumulative knowledge of these quantities over all possible risk mechanisms, an assessment can be made of the likelihood of various outcomes. These can be compared to a business's risk posture to determine if the plan is good or if alternates should be sought.

A risk analysis for METOC would similarly require an assessment of the probability of unpredicted harmful conditions weighted by an assessment of the impact of those conditions on the goal (e.g., determination of the impact of mission failure and/or loss of life). Together, these can be defined as the cost of uncertainty.

The Idea of a Cost of Uncertainty

The environment affects all military operations, sometimes simply by introducing inconvenience but other times in a show-stopping role that can change history. Thus, it is no surprise that the environment has played an important role in military conflicts throughout the ages. For example, the harsh Russian winters repeatedly doomed invading armies. At sea, entire fleets have been lost due to foul weather. More recently, the failure of the rescue mission of the hostages in Iran had enormous costs, holding captive U.S. foreign policy for a year and contributing to the demise of the Carter presidency. The successful effort to rescue U.S. citizens on the island of Grenada in 1983 was marred by the loss of four Navy SEAL team members involved in a night jump into coastal waters during marginal weather conditions. It has been estimated that 75 percent of naval special warfare missions are severely degraded due to METOC problems (LCDR Bruce Morris, personal communication).

The Goal of the METOC Enterprise

It is unlikely that we will be able to control the weather in the foreseeable future; thus, our goal is instead to know the current state of the environment and make useful predictions of its future state. The environment affects both sides of a conflict. Difficult conditions can be an impediment or at times an advantage.

The tactical goal is to minimize risk while maximizing the possible advantage of the environment. The METOC goal is to facilitate that objective through perfect knowledge of the environment.

In a world of perfect environmental knowledge, decisions made by leaders in mission planning could be perfect.[5] However, constraints of both a physical nature and a fiscal nature will always limit our ability to know and predict natural systems. It is the introduction of uncertainty due to environmental unknowns that has risk and introduces cost. Thus, reduction of this uncertainty is the goal of the naval METOC enterprise, and it is the cost of the uncertainty that provides a gauge for the value of METOC knowledge.

It is a central hypothesis of this study that the goal of the naval METOC enterprise is the reduction of uncertainty due to environmental factors in mission planning and operations. Thus, the optimum investment strategy is that which provides the largest reduction in the cost of uncertainty for the smallest investment cost. An important consequence of this hypothesis is the need for the naval METOC community to embrace the concept of uncertainty and uncertainty reduction as fundamental to METOC products.

SUMMARY

The naval METOC enterprise is a complex system of platforms, personnel, and computer systems designed to support operations carried out by the U.S. Navy and Marine Corps by producing high-quality tailored environmental information products. The intended consumers of this information include decision-makers facing a variety of complex choices, some of which may be significantly affected by environmental processes operating at a variety of temporal and spatial scales. There is an apparent lack of a clear performance metric, including a robust understanding of how platforms and personnel are affected by environmental processes. Valid and quantifiable feedback from warfighters and other operators, while difficult to obtain, is needed if objective criteria for data acquisition are to be established. Limited funding, limited time, and the rapidly evolving nature of naval and expeditionary warfare make sound decisions regarding data acquisition and dissemination a high priority in the naval METOC community. Establishing the value (dollar amount) of environmental information for military decision-making may seem an impossible task given the complexities of military operations and the uncertainties of enemy intent. The value of such an exercise would be substantial, just as understanding the economic costs of natural disasters helps define debates over budget priorities for emergency services. The first step in understanding the value of environmental information lies in understanding its present use in various naval operations.

---

[5]Defined here in a METOC sense. Parallel arguments can be made to place value on intelligence information.

3

Nature of the Problem: Sources and Limitations of METOC Knowledge

> This chapter points out:
>
> - the technical challenges encountered in characterizing the environment in support of military operations;
> - the limits of current scientific understanding and how they affect the ability of the meteorological and oceanographic (METOC) enterprise to fully meet the needs of U.S. Naval Forces (the U.S. Navy and Marine Corps); and
> - how uncertainty inherent in complex natural systems, coupled with that associated with the limitations of technical systems and scientific understanding, can translate into operational challenges for U.S. Naval Forces.

The discussion in the preceding chapter identified a large number of environmental parameters of interest to U.S. Naval Forces. These parameters affect, among other things, visibility, ocean acoustics, and the propagation of electromagnetic waves. If some determination of required environmental parameters can be made for any particular mission, the next question is "Can the values of the required parameters be determined at the particular time and place of the mission and to the requisite accuracy?" This question has no simple answer because, for any particular parameter, the answer is a convolution of many factors. These include:

- The importance of the parameter to the specific concerns of a given naval mission, such as intelligence gathering, use of certain weapons systems, and aircraft operations. Some of the parameters may be mission critical; others may be of much less importance.
- The accuracy and precision to which the parameter value must be known. In many cases, the parameter may have a critical threshold for making a go/no go decision. It may be sufficient to know the value with only modest accuracy when the value is far from the threshold but necessary to know it to high accuracy near the threshold.
- The spatial and temporal variability of the parameter. In some cases, an average value over relatively large areas and long times may be sufficient; in other cases, pinpoint accuracy at specific times and places may be required.
- The relative uncertainty of the predicted value. In practice, the decision to proceed with a particular naval mission depends on many factors, one of which is the environment. The role of the environmental factor in the ultimate decision process is determined in part by the certainty of knowledge of particular key parameters. The more certain the value, the more probable its use in decision-making.

One approach to dissecting this problem into manageable components is to sort and organize parameters by spatial and temporal scales and associated predictability. The following section discusses the relationship of mission time lines to physical knowledge. The next section focuses on predictability and model scale. The final section outlines data availability and its relevance to mission timescales.

IMPRINT OF PHYSICS ON MISSION TIME LINE

Figure 3-1 presents a schematic space-time diagram illustrating the spatial and temporal scales of variability of many environmental systems. Of particular importance in evaluating this diagram is development of an objective method for understanding present and future capabilities for predicting environmental conditions across different spatial and temporal scales. For example, at the largest spatial (global) and temporal scale (decadal), climatological data can be treated statistically to arrive at estimates of the state of the global environmental system for any interval of the present year. These statistics are largely probabilistic functions defining the likelihood that the system will be found to lie within defined bounds. At this scale of observation, present-day predictive skill is reasonably well developed. However, as the spatial and temporal scales of processes become smaller, the influence of nonlinearities in each environmental system becomes larger, resulting in greater uncertainty in the ability to predict the state of the environmental system. At the smallest time- and space scales, nonlinearities result in environmental processes that are dominated by stochastic processes such that

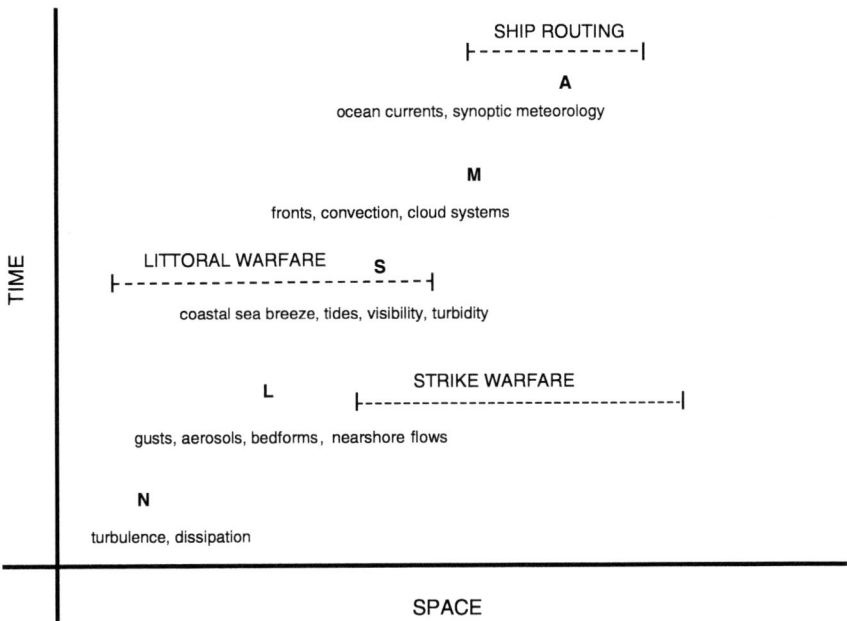

FIGURE 3-1 Schematic diagram showing the dominant spatial and temporal scales of physical motion in the ocean and atmosphere. The letters above the group of motions indicate the predictive skill associated with each: N—none, L—low, S—some, M—moderate, and A—acceptable. The overlap of some naval mission areas with specific ranges of time- and space-scales is shown. The range of model types used to simulate, forecast, and predict the various motions is shown along the bottom of the figure.

prediction is futile, and present-day capabilities rely on observation rather than modeling to describe the system.

The primary issue related to these scaling prediction phenomena concerns identifying the thresholds where reduced predictive power should be replaced with observational descriptions. A significant component of naval research into these areas would be beneficial because not only would it help establish additional constraints on understanding environmental information but it would also help shape assessment of the need for sensors and sensor arrays.

The range of physical processes in the ocean and atmosphere and the interactions between them are fundamental in determining the physical structure that is encountered at any time and location. Space-time diagrams (e.g., Stommel, 1963) provide a useful way to view the possible time- and space scales over which oceanographic and atmospheric processes operate. These scales typically cover many orders of magnitude and range from dissipative processes, which operate over short time- and space scales, to scales associated with large-scale ocean currents and atmospheric systems.

The various time- and space scales associated with physical processes are not, however, independent of each other. Rather, these scales form a cascade through which information is transferred from large to small scales and vice versa. Consequently, perturbations introduced at one scale can potentially affect processes at a large number of smaller and larger scales.

The nonlinear interactions that occur between processes at different scales make predicting future states of the ocean and atmosphere a challenge. Atmospheric and oceanic numerical circulation models are constructed to include the specific dynamics that are related to particular ranges of time- and space scales. Global models treat processes at the largest scales but are forced by limitations in computer resources to parameterize all processes at the intermediate to small scale. Regional-scale models introduce processes that occur at larger scales via boundary conditions, such as the effect of large-scale ocean currents on regional flows, but still include the smaller-scale processes via parameterizations, such as turbulence closure terms in circulation models. High-resolution models include explicit representations of many small-scale processes but are often of limited value in a predictive sense because there are insufficient data to initialize the model accurately.

Due to the broad range of interacting time- and space scales, oceanic and atmospheric circulation models inherently provide a filter to the range of dynamics being simulated in any particular model. This characteristic (and limitation) has several important consequences. First, the predictive capability of any model is related to multiple factors, among them the scales that are being resolved in the model, the persistence of those scales in the ocean or atmosphere, and the quality of the initial and boundary condition data. Second, because of the nature of the cascade process and limits in our understanding of physical processes, predictive skill is generally better at larger spatial scales and the associated timescales. Third, the urgent need for initial and boundary condition information provides a strong rationale for developing data assimilative modeling efforts and acquiring data for input to these models. Without such efforts it is unrealistic to expect oceanic and atmospheric circulation models to have good predictive skill over a wide range of time- and space scales.

Areas of Acceptable Predictive Skill

Because of the wide range of oceanic and atmospheric scales of interest for various applications and the equally broad range of predictive skills, it is useful to introduce the concept of acceptable predictive skill. This concept is an attempt to merge our understanding of the predictive skill of a particular model with the needs of a particular application. For example, consider the problem of predicting hurricane landfall along the Atlantic coast. Currently, atmospheric models

Radio communication with warfighters deployed inland can be adversely affected by local meteorological conditions. Phenomena such as electromagnetic ducting may allow signals to carry too far, allowing detection by opposing forces, or can disrupt signals sufficiently to pose tactical problems (Photo courtesy of the U.S. Navy).

exhibit considerable predictive skill on the 24- to 48-hour timescale but are much less skillful on timescales greater than 72 hours. If landfall predictions on the order of 24 to 48 hours are supporting decisions regarding when to put the fleet to sea, the current situation represents acceptable predictive skill *for the Navy* and no additional effort is required. If a longer warning time is needed, the Navy must consider what research, data, and financial resources are needed to make forecasts reliable on that timescale.

Some ranges of time- and space scales can be simulated with acceptable predictive skill for Navy applications. For oceanic systems these include the large-scale ocean current structure and long-term climate variability. For example, considerable progress has been made on forecasting the El Niño-Southern Oscillation (ENSO) in the tropical Pacific Ocean with lead times of 6 to 12 months. Intercomparisons of the models used for ENSO predictions show that the forecasts are considerably better than those obtained by simply using persistence of features to determine future conditions (Latif et al., 1998). Although the ENSO models have proven predictive skill, they still need improvement and development, especially in regard to specification of forcing fields, model initialization, and data assimilation capability (Stockdale et al., 1998).

Similarly, simulation of large-scale ocean currents, such as those obtained from the Ocean Circulation and Climate Advanced Modeling project (Saunders et al., 1999; Webb et al., 1998), show good agreement with observations. A recent test and intercomparison of simulations of the North Atlantic obtained from five numerical circulation models showed that all were able to simulate the large-scale current structure with a fair degree of realism (Chassignet et al., 2000; Haidvogel et al., 2000).

Global weather forecasts have shown substantial improvement over the past two decades, which has been quantitatively demonstrated by a variety of studies of model skill scores. The evolution of weather patterns can be predicted with skill on timescales of five to seven days, depending to some extent on season and latitude zone. Forecasts of severe weather events have shown a corresponding increase in skill. The use of satellite data in data assimilation models has brought significant improvement to ocean weather forecasts, particularly in the tropics and southern hemisphere where conventional meteorological data are sparse. These global models typically provide reliable information on the scale of 100 km and greater.

Areas of Moderate Predictive Skill

During the past decade much effort has gone into the development of oceanic numerical circulation models for simulation and prediction of regional-scale flows (e.g., Haidvogel and Beckmann, 1998). For some areas, regional circulation models have undergone testing in terms of their ability to forecast such features as ocean fronts (e.g., Robinson et al., 1996).

Most forecasting centers employ regional atmospheric models to provide higher-resolution forecasts over regions of interest. These models run nested into a global model (i.e., the global model provides the boundary conditions at the lateral edges of the nested regional model). Because of problems associated with the propagation of information from the boundary, these regional models typically provide skillful predictions on the order of 48 hours at resolutions of 20 to 40 km. The higher resolution captures mesoscale effects such as terrain variability and lake effects with greater fidelity than can be achieved with the global models. As computer capability increases, the tendency is to replace these regional models with global models of the equivalent resolution.

Areas of Some Predictive Skill

Because of the stochastic nature of small-scale atmospheric motions and the spatial scale of cloud processes, atmospheric models are limited in their ability to forecast the strength and timing of many severe weather events. A regional forecast model may be quite capable of predicting the conditions conducive to severe thunderstorm development and associated weather, such as hail or tornados, but

An Explosive Ordnance Disposal Technician, assigned to Explosive Ordnance Disposal Mobile Unit Five (EOD MU-5), leaps from a CH-46 "Sea Knight" helicopter during a static line water parachute jump over Apra Harbor, Guam. Atmospheric and oceanographic conditions are an important factor in ensuring the safety of personnel involved in such operations (Photo courtesy of the U.S. Navy).

cannot predict the actual location or timing of these events with great certainty. Sea breezes, which depend critically on relatively small-scale thermal gradients and near-shore terrain, are another phenomenon that can be predicted by models in a broad sense but can have significant errors in locally predicted values. In many cases, experienced forecasters who are familiar with the performance of the typical forecasting models can correct these model errors based on their own knowledge.

Similarly, because of the scale and variability of surface-atmosphere interactions, boundary layer development and modification are predicted less well than free tropospheric properties. This affects model predictions of a variety of smaller-scale processes, including boundary layer cloud, radiation fog, and local precipitation. Aerosol production at the sea surface and aerosol lofting and transport are also affected negatively by this problem.

Areas of Low or No Predictive Skill

A variety of high-resolution atmospheric models are used to study cloud development and boundary layer structure. These models typically incorporate very detailed representations of both dynamics and cloud processes. Their output, however, is largely statistical in nature and not predictive in a deterministic sense. This arises from two principal causes. First, there are no observation grids that supply data with sufficient temporal and spatial resolution to reliably initialize and force the model. Second, turbulence and convection are inherently stochastic processes that introduce randomness into any simulation. Thus, it is unlikely that we will ever be able to predict the exact nature of boundary layer structure and cloud formation on small time- and space scales.

DATA ACQUISITION

Environmental data used by the Navy for various purposes can come from many sources. The Naval Oceanographic Office (NAVOCEANO) operates a fleet of oceanographic survey ships to collect information from the seafloor, ocean surface, and water column. The TAGS50 and TAGS60 class ships provide multipurpose oceanographic capabilities in coastal and deep ocean areas, including physical, chemical, and biological oceanography; multidisciplinary environmental investigations; ocean engineering and marine acoustics; marine geology and geophysics; and bathymetric, gravimetric, and magnetometric surveying. Much of the data collected by this fleet are permanently archived in databases maintained by NAVOCEANO. Many of these data are compiled and written to CD-ROM on an as-needed basis to accompany OA[1] officers as they embark on tours of duty. Typical archival datasets carried onboard fleet assets in this way might be regional tide/lunar tables covering the operating area, archival bathymetric data, digital nautical charts and digital terrain models of coastal regions or islands, and ocean/water column and atmospheric climatologies.

Data systems that are being developed for the METOC community are focused on access to meteorological and oceanographic data and products generated by the Navy and/or passed through naval [and in some cases other Department of Defense (DOD)] data centers. There is little to no consideration in the development of these systems for direct access to data that are generated and held outside DOD despite the use of such data sources via standard Web browsers by METOC personnel in recent naval operations. Historically, this focus has made sense in that access to near real-time data and/or data products has been provided by a small number of government data centers, greatly facilitating the task of providing access to these data via a single system. There is, however, a clear

---

[1]The acronym "OA" refers to the military designation "Operations, Aerography" and dates back to a period in naval history when METOC personnel were rated as aerographers.

trend toward increased development of and open access to real-time data and data products in the commercial and non-DOD research sectors (this is explored more fully in Chapter 4).

A large area of the world surface and ocean environment, however, remains sparsely instrumented, so in situ data needed to initialize and run regional models may not be available. Strategies to develop sensors or sensor arrays for collecting data in these regions will need to be developed. The U.S. Navy's remote sensing program focuses on tactically significant information in the littoral regions. Environmental sensors on existing satellites are being exploited to the greatest extent possible, and new capabilities, such as those offered by WinSat and hyperspectral imagers, are under development. Environmental sensors on airbreathers, surface ships, and undersea systems appear to be underutilized, as are environmental data that can be extracted from sensors being deployed for other-than-environmental missions.

IMPEDANCE MISMATCH FROM DATA SOURCE TO WEAPONS SYSTEM: AN EXAMPLE FROM UNDERSEA WARFARE

Environmental information for undersea warfare primarily concerns understanding how the ocean impacts acoustic propagation and the design and performance of Navy sonar systems. Sonars are usually divided into passive, which exploits radiated sound from a source, or active, which uses echoes to resolve the range and Doppler shift of a source. Support for both passive and active antisubmarine warfare (ASW) missions dominated funding for environmental characterization during the Cold War. It still does, but with the emphasis now on littoral waters instead of the historic focus on deep water. Moreover, the shift of Navy missions toward the littoral has placed increased emphasis on mine countermeasure (MCM) systems as well as torpedoes.

There are many different platforms and missions, so an overview is useful for organizing the discussion. The platforms specifically relevant to undersea warfare can be divided as follows: submarines, surface ships, aircraft, and surveillance systems. Furthermore, the missions of each can be listed: for submarines, self-defense, ASW, antisurface warfare, ISR (intelligence, surveillance, and reconnaissance), CVBG (carrier battle group escort), Naval Special Warfare and deployments, clearing contested areas and commercial shipping routes, and forward strike (cruise missile launching). The undersea warfare for surface ships, typically cruisers and destroyers, concerns self-defense, ASW, and CVBG escort. Aircraft, both maritime patrol aircraft and helicopters, are almost solely concerned with ASW operations. Similarly, the Integrated Undersea Surveillance System, including fixed assets such as the remnants of the SOSUS system, the ADS advanced deployable system, the fixed deployable system, and SURTASS (Surveillance Towed Array System) are concerned with both ASW and commercial ship surveillance.

The emphasis in this section is on environmental support for ASW-related operations; nevertheless, MCM and weapons systems are also important Navy missions. In comparison to ASW, these sonar systems do not use a lot of environmental information with the exception of seabed characterizations for reverberation predictions. The complexities of the littoral will require more environmental support to improve the performance of these systems.

Scales for Environmental Characterization

Spatial and temporal scales are very important aspects of environmental data needed in undersea warfare. High-resolution geoacoustic data acquired over large areas are costly and time consuming. GDEM products represent a long-term NAVOCEANO effort for seafloor characterization. Uncertainties about the seafloor are a dominant issue in transmission loss calculations. The water column and sea surface are dynamic processes whose prediction has also been a long-term objective of NAVOCEANO. The MODAS effort fuses archival databases, satellite altimetric data, circulation models, and in situ data for these predictions. The needed scales for accurate predictions are fundamental and still very much of a research issue. The often-stated assertion is that lack of high-resolution environmental data is the major deficiency for predicting the performance of a sonar. While this overstates the current capability in acoustic modeling, the statement misses an important point. One really wants to identify the spatial and temporal scales where robust predictions are possible.

Another important aspect of the use of environmental information is the time and spatial scales for predictions. Atlases, databases, and climatology are important for general assessments, mission profiles, and the regional environment. Tactical needs are much more specific, but they can exploit real-time in situ measurements such as sound speed profiles, ambient noise distribution, and areas generating significant backscatter or acoustic reverberations. This context is the one that stresses current prediction tools. Finally, postanalysis and evaluation of new algorithms use all available data. Without the urgency of real-time decision-making, tradeoffs involving environmental uncertainties can be done.

Environmental Components of Undersea Warfare

The requirements for shallow-water (i.e., bottom-limited) acoustic predictions transcend predicting propagation loss and reverberations; the data must be presented to the operators in a usable manner. Since today's sonars have sufficient power to be able to project sound well beyond the first bottom interaction, multiple zones of potentially useful ensonification occur with each ping. The challenge is to present that to the operator in a way that enhances exploitation.

The concept presented in this chapter of the stochastic nature of the environment and its characterization in models, is especially valid when it comes to

The nuclear-powered Los Angeles class submarine USS Key West (SSN 722) conducts surface operations. The boat is part of the aircraft carrier USS Constellation (CV 64) battle group en route to the Arabian Gulf to enforce no-fly zones and monitor shipping to and from the region. Understanding how ocean and seafloor conditions affect acoustic properties can provide important tactical advantages to U.S. Navy submarines engaged in protecting surface vessels or other assets threatened by enemy submarines (Photo courtesy of the U.S. Navy).

acoustic propagation and prediction. Ping-to-ping variations are well documented and render all conversional prediction systems, which use signal excess as a measure of probability of detection, invalid. Note: The ASW world is already used to dealing with predictions stochastically since it is "probability of detection" and not "certainty of detection" that is dealt with. It is entirely possible that this nuance is lost on many operators who take terms such as "range of the day" literally. The same is true in weather forecasting, where such terms as "strike probability" and "chance of rain" are often misunderstood. Since these types of uncertainties are the norm in natural systems, emphasis should be on understanding and exploiting this characteristic.

The components of the sonar equation provide a useful guide to the way environmental information is used in undersea warfare. For passive systems:

$$SE = SL - TL - NL + AG + SPg - RD$$

where SE is the signal excess, SL the source level, TL the transmission loss, NL the ambient noise level, AG the spatial array gain, SPg the signal processing gain, and RD the recognition differential. For active systems in a reverberation-limited environment:

$$SE = SL - 2*TL + TS - RL + AG + SPg - RD$$

where SL is now the power of the active source, TS the target strength, and RL the reverberation level. The environmental components are TL, NL, and RL and to a lesser extent AG and SPg.

Prediction of the effects of the environment on transmission loss has long been a goal for Navy systems. In its simplest context this led to the "range of day" figure of merit. The problem now, especially in shallow water, is recognized to be more complex than just a single number. A number of propagation codes exploiting Navy databases such as SFMPL and PCIMAT are now available and can be run onboard. Most of these use a parabolic equation method that is a narrow-band, range-dependent code. The water column, surface, and bottom can impact TL, so the problem can be quite complex. One of the important goals of NAVOCEANO is to provide the environmental databases and prediction tools for accurate predictions of TL. The claim is an accuracy of ±2 dB when in situ sound speed is well known and good bottom models are available. Operationally, such performance is not routinely achieved.

Predictions of TL for deep-water RR (refracted/refracted) and RSR (refracted/surface reflected) are easier because the bottom is not involved. These paths were very important for long-range submarine detection during the Cold War. Significant problems appear when there is a significant number of bottom interacting paths (i.e., not enough excess depth). For deep water and short ranges there are usually two dominant paths between a target and a submarine array—

direct paths refracted and/or reflected from the surface and bottom bounce paths. Direct paths are relatively easy to model for TL. The bottom bounce paths are very difficult to model and the uncertainty is very high. This leads to problems for predicting the performance of an active bottom bounce sonar. For a submarine, changes in bottom loss manifest themselves as the part of the TL plot versus range where there is cylindrical spreading. Large changes in range lead to relatively small changes in TL, so there is a lot of uncertainty about counterdetection ranges. Currently, TL predictions when there are bottom interacting paths are not considered very reliable.

Reliable TL predictions are not available for littoral shallow waters. There are many examples where the differences among model predictions and calibrated experimental data differ more than 10 dB. Shallow-water environments are currently the dominant regions for naval operations; therefore, this is a significant problem. Horizontal propagation in shallow water is dominated by the seafloor. Steep paths, or high-order modes, above the critical angle are rapidly attenuated. The near grazing paths, or low-order modes, propagate with a complicated dependence on the bottom. Separating the effects of attenuation due to absorption, scattering, and shear conversion losses is difficult. The parabolic equation approach does not account for variability due to oceanographic processes and the seabed. Including variability in prediction models is an ongoing research issue for TL. Overall, implementing the correct geoacoustic model of the seabed in shallow water for TL still has many unresolved problems.

The current NAVOCEANO approach uses broadband measurements on a line with increasing separation between source and receiver. Stratigraphic models are based on reflection/refraction profiles. Genetic algorithms are then used to tune the layer parameters of a geoacoustic model such that there is less than ± 2 dB frequency averaged difference between the parabolic equation and 1/3 octaves of the broadband data. The fundamental issue is how robust this approach is and how detailed must the databases be to lead to accurate TL predictions.

Ambient Noise

Beyond TL the ambient noise environment is the second major factor in ASW performance. Ambient noise can be divided into two major components—the diffuse ambient background and strong discrete interference from shipping. The diffuse component represents unresolvable distant shipping and wind/wave surface-generated noise. "Discretes" are signals from high shipping noise, geophysical seismic exploration, and biological sources. (Vocalizations from whales and seismics can be the major contributor to ambient noise in some regions depending on the season.) The often-termed "clutter" leads to "stripes" on bearing time displays since they are typically at long ranges and have small bearing rates. Because of their high levels, these sources do interfere with detections, especially if a target also has a low bearing rate. Ambient noise databases are maintained (e.g., the HITS database for commercial shipping is often used for modeling discrete shipping noise, while the ANDES [Ambient Noise Directionality Estimation System] and DANES [Directional Ambient Noise Estimation System] models are used for the deep-water diffuse noise levels).

Ambient noise in the littoral is a much more complicated issue. The ranges are typically shorter, so fishing fleet and other coastal traffic become strong contributors. While there are seasonal trends for fishing, databases comparable to HITS are not available. Moreover, the complexities of littoral acoustic propagation lead to modeling problems for prediction models.

Ambient noise is important for several aspects in sonar performance predictions. First, one obviously wants to understand regional levels, so as not to operate where a target is in the same sector as high levels of ambient noise—quiet beam sectors are important. Next, in deep water ambient noise often has a vertical distribution with a noise notch at near grazing angles that can be exploited. At any time there are many discrete clutter lines whose exact bearings cannot be predicted; nevertheless, prediction of their distribution is important. Finally, many of the advanced beamformers such as those used for the evolving ARCI [Acoustic Rapid COTS (commercial-of-the-shelf) Insertion] systems are based on noise cancellation algorithms whose performance strongly depends on the distribution of the discrete components of the ambient noise field to improve the array gain term in the sonar equation.

Ambient noise levels enter the sonar equation just as TL does; however, models and databases especially for the littoral have not kept pace and have not received a comparable level of effort.

Reverberation

Active sonar systems usually operate in a reverberation-limited environment. Increasing the power in such an environment not only increases the echo return from a target but also the level of interfering reflectors such as the seabed, sea surface, and fish. The only way to improve performance in a reverberation-limited environment is by signal design. Reverberation from the bottom is by far the most important in most Navy sonars.

Reverberation is characterized in terms of the scattering strength or rms level per unit area. The scattering strengths as a function of frequency and angle of incidence of many regions have been measured. Data from many important operational areas, however, are not available. The SABLE [Sonar Active Bottom Loss Estimate] program addresses reverberation characterization by making "through the sonar" measurements with operational sonars.

Reverberation in active systems is the equivalent of ambient noise for passive ones. Scattering strength is a very simple description that misses many of the important issues in predicting performance. Discrete reverberation appears as "highlights" and possible false detects. There is also significant local variability within a geological province. At low frequencies and low sediment cover, detailed bathymetry enables good predictions. At higher frequencies and sedimented areas typical of littoral operations, however, this becomes more problematic since the wavelengths are shorter and geoacoustic models are not available. Moreover, the acoustics of some of these models are often in doubt because of buried features.

An important question in ASW concerns the limits of passive systems leaving active systems as the alternative. While reverberation has received significant attention throughout the development of the SQS-53, LFA (Low-Frequency Acoustic), and LFAA (Low-Frequency Active Adjunct) systems, existing databases are not adequate for good predictions and much remains in understanding the geologic control on seafloor acoustics.

Platform-Specific Issues

Submarines

The undersea environment dominates the operation of a submarine. Characterizing the in situ sound speed profile and local bathymetry is crucial for predicting the performance of a submarine sonar and the ability to fulfill the self-

The guided-missile frigate USS McClusky (FFG 41) conducts a tight navigational drill. Frigates fulfill a "Protection of Shipping" mission as anti-submarine warfare combatants for amphibious expeditionary forces, underway replenishment groups, and merchant convoys (Photo courtesy of the U.S. Navy).

defense, ASW, and CVBG escort roles and for clearing contested areas and shipping lanes. A submarine uses several passive arrays for detection and tracking. These include the sphere (BQQ-5) operating from 1.5 to 7.5 kHz, conformal (hull) arrays, several towed arrays (TB-16, TB-23, and TB-29) operating below 1 kHz, and on some boats the WAA (wide aperture array) for wide-aperture ranging.

Detection ranges for undersea warfare have decreased significantly to the order of tens of kilometers because of adversary stealth and quieting plus the introduction of quiet diesels operating on batteries. (This problem becomes even worse with the introduction of air-independent propulsion systems.) No longer can tens to hundreds of kilometer ranges be maintained with lots of time to maneuver. With such short detection ranges many issues become more significant: (1) ship safety (avoid a collision); (2) counterdetection (crucial for self-defense); and (3) rapid and tight maneuvers to maintain tactical advantage, which severely degrades towed array performance. All these issues place a premium on accurate predictions of transmission loss and the uncertainty associated with them. It should be noted that the shorter ranges imply less reliance on long-range refracted-reflected and refracted-surface reflected paths and more on bottom-interacting paths. The important environmental issues concern the available databases and tactical decision aids (TDAs) to use them.

For other missions, such as intelligence, surveillance, and reconnaissance (ISR) operations, special forces insertions, and forward strike, environmental information is crucial for navigation, safety, and maintaining stealth.

Surface Ships

Surface ships employ both active and passive systems. A forward cylindrical array (SQS-53) operates in both the active and passive modes. Towed arrays (SQR-19) are used as well. Surface ships have most of the same missions as submarines. They have the advantage that stealth is not an issue; thus, active sonar can be used, although the absence of stealth is a disadvantage for self-defense. All the same issues for submarine passive sonar are present in surface ship passive sonars, so the need for databases and TDAs is much the same. The active systems must contend with reverberation that dominates system performance in littoral waters. Unfortunately, good databases and TDAs for reverberation predictions are not available in many operationally significant regions.

Air ASW

Air ASW involves deploying hydrophones or arrays of hydrophones from MAP or helicopters. Generally, they are short-range systems for localization. Over time these systems have evolved from single-hydrophone sonobuoys to directional sonobuoys (DIFARS) using gradient sensors, vertical line arrays (VLADS), and horizontal arrays (ADAR) for passive systems. Active systems have evolved from single-sonobuoy sources and receivers to long-range systems integrated with surface and submarine sonars.

Surveillance

For many years the SOSUS (Sound Surveillance System) network of cabled horizontal arrays on the seafloor was the key component for long-range ASW. An extensive amount of environmental information was used to model the performance of these arrays. Most of our knowledge of long-range acoustic propagation is the result of support for SOSUS. Much of the SURTASS (Surveillance Towed Array System) system was abandoned at the end of the Cold War. Moreover, detection ranges from SOSUS decreased significantly as the result of quieting.

SOSUS has now been been supplemented by SURTASS, a system of large-aperture towed arrays. In addition, the FDS, a large network of fixed seafloor arrays, and ADS, a rapidly deployable network of seafloor arrays, have been deployed and/or tested. Recently, SURTASS has been augmented with an active component forming the LFAA (Low Frequency Active Adjunct) system. The important issue is that the same requirements for accurate predictions remain—that is, good TL estimates and robust ambient noise models.

The LFAA, and more generally the use of all operational active systems, recently added another dimension. The need to predict the impact on marine mammals has become an emotional public debate, so much so that the operation of an active system is called into question. Prediction levels received from an active source by a mammal and background ambient noise levels are major issues in the debate.

SUMMARY

The spatial and temporal scales of various environmental processes may have a profound effect on efforts to predict future states. Areas where inadequate predictive skill is of concern require greater numbers and more recent observations, to either support data assimilation or supplant predictions altogether. Determining what level of accuracy is needed in predicting future states of some environmental phenomena is severely limited by a clear understanding of the sensitivity of platforms and personnel to various environmental conditions. Thus, before rigorous efforts at setting data acquisition priorities can be completed, a fuller account of the impact of various processes on naval operations must be acquired. The General Requirements Database is a good first step, but greater and more clearly defendable establishment of critical thresholds, including compound effects from multiple environmental processes, is needed. With this information in hand, efforts to weigh the benefit of additional information against the cost of acquiring it could be undertaken.

4

Improving Environmental Information by Reducing Uncertainty

This chapter points out that:

- uncertainties about environmental conditions, and information describing them, can have tactical significance;
- the most cost-effective method for reducing the impact of uncertainty may vary with different types of environmental information or its intended use;
- when additional observations are key to making significant improvements in the quality of environmental information (i.e., reducing uncertainty to acceptable levels), the most cost-effective means should be explored first; and
- delivering a sensor to a denied location of interest may represent a significant component of the cost of collecting observations. Thus, using sensors or sensing systems employed for intelligence, surveillance, or reconnaissance purposes may present an attractive option so long as the primary mission is not significantly impaired.

As discussed in the previous chapters, the meteorological and oceanographic (METOC) enterprise can be viewed as an organized effort to provide information useful to naval operations about the current and future state of the environment. This process is not perfect, but instead introduces errors that can be characterized by associated uncertainties. Thus, environmental information either explicitly or implicitly contains uncertainty that is inherit in the stochastic nature of environmental processes or that is introduced by imperfect sampling or numerical calculations using data from imperfect sampling. Understanding and living with these uncertainties, especially understanding when it is important to reduce them and how, are a primary focus of this report.

Webster's defines uncertainty as (1) the quality or state of being uncertain and (2) something that is uncertain. Thus, to the layperson, such predictions might be viewed as efforts to reduce the uncertainty associated with the nature of the terrain just over the horizon or forthcoming weather conditions. In statistics, the physical sciences, and other technical fields, however, the term *uncertainty* holds several specific definitions that can be expressed mathematically.

According to Ferson and Ginzburg (1996), there are two basic kinds of uncertainty. The first kind, *objective uncertainty*, arises from variability in the underlying stochastic system. The second kind, *subjective (epistemic) uncertainty*, results from incomplete knowledge of a system. Objective uncertainty cannot be eliminated from a prediction regardless of the number of previous observations of the state of a particular environmental condition (e.g., no matter how many thousand previous waves are observed, the height of the next incoming wave cannot be predicted without some uncertainty). Data collection and research into the environmental processes shaping future states can be used to understand and, to some degree, reduce uncertainty, but not without some additional cost (not just in terms of resource expenditure but also in tactical advantage as some efforts may alert opponents to a pending military operation). Probability theory, and other approaches, may provide methods appropriate for projecting random variability through the calculations that result in quantitative predictions of future states.

TACTICAL IMPLICATIONS OF UNCERTAINTY

Military decisionmakers rarely deal with uncertainty in a formal, statistical manner. Nevertheless, many decisions can be examined in such terms. In situations where uncertainty about environmental conditions is straightforward—that is, when uncertainty about current conditions will not impact predictions of future states—commonsense approaches are very effective and are typical of sound decisionmaking. Such examples are common in military situations. By examining such tactical decisions involving environmental information one can gain insight into how more complex situations may be dealt with.

In the following example from mine warfare (modified from an example presented at the Symposium on Oceanography and Mine Warfare organized by the National Academies' Ocean Studies Board, in Corpus Christi, Texas, in September 1998; see National Research Council, 2000), coalition forces have been tasked with securing a seaport to expedite pacification of a hostile coastal nation (see Figure 4-1). The geology of two offshore islands lends itself to the natural development of many small boulders, which occur on the adjacent sea-

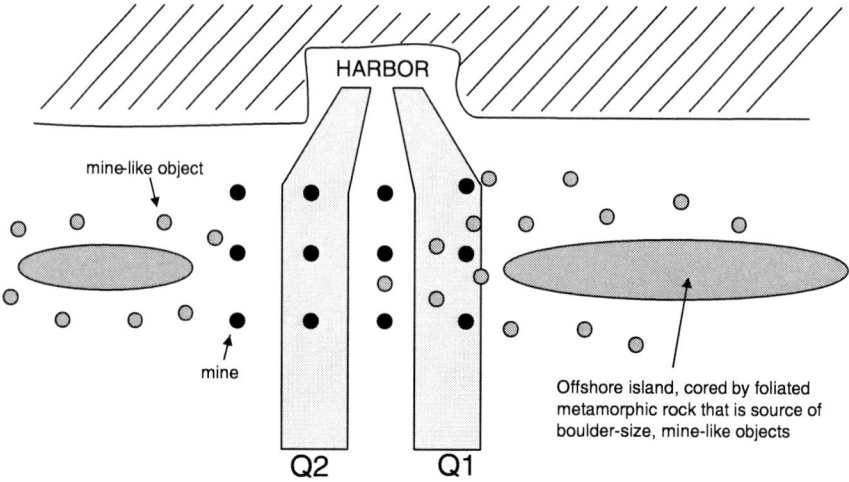

FIGURE 4-1 Cartoon of the distribution of mine-like objects and mines offshore of a foreign harbor. Since resources must be expended to determine whether each object is or is not a mine, the cost of uncertainty associated with Q-route 1 is much greater than that for Q-route 2. Thus, Q2 is the preferred route.

floor in large numbers. Typical seafloor mapping techniques can resolve objects of this size, referred to as mine-like objects, but distinguishing mines from other mine-like objects requires the use of unmanned underwater vehicles (UUVs), EOD (explosive ordinance disposal) divers, or other means. Because this is time-consuming and often dangerous work, practical decisionmaking simply identifies a shipping lane (referred to by the U.S. Navy as a Q-route) with the fewest number of mine-like objects. Thus, in Figure 4-1 the uncertainty associated with the nature of mine-like objects (are they mines or not?) is greater along Q1 than along Q2. Q2 becomes the obvious choice, and appropriate mine countermeasures would be applied as needed. No sophisticated or rigorous effort to quantify uncertainty is needed. It is interesting to note, however, that if the opposing forces understood the nature of the seafloor and incorporated consideration of uncertainty even in an informal manner into their mine-laying plans, a more efficient use of the limited number of mines available could have been achieved.

As shown in Figure 4-2, by concentrating mine laying in areas with a lower density of naturally occurring mine-like objects, opposing forces would eliminate the most obvious Q-route, forcing coalition forces to expand additional resources to identify and eliminate mines. This practical tactical application may seem implausible, as Q1 is not actually protected by mines. However, if the com-

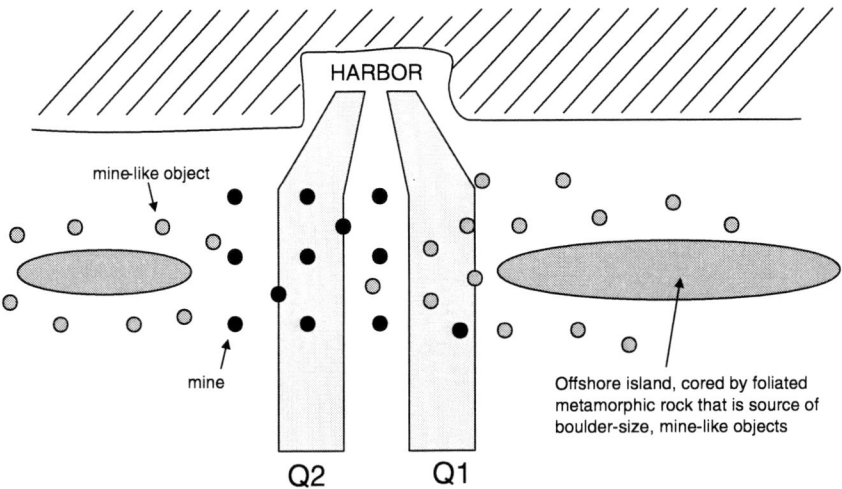

FIGURE 4-2 Cartoon of the distribution of mine-like objects and mines offshore of a foreign harbor. Since resources must be expended to determine whether each object is or is not a mine, the cost of uncertainty associated with either possible route is roughly equal.

mander of the opposing force is convinced that coalition forces will delay efforts to secure the harbor until risk to transport ships can be held to a minimum, mine laying becomes more of a delaying tactic and the distribution of mines in Figure 4-2 is a better tactical decision. This example points out several aspects of dealing with uncertainty that can be explored in more rigorous ways in more complex examples. Foremost among these is the decision to balance the benefit of reducing uncertainty against the cost. In more complex decisionmaking scenarios involving environmental uncertainty, cost versus benefit will become a more important factor.

Propagation of Uncertainty in Complex Systems

In the previous example, uncertain knowledge of the true nature of mine-like objects is easily quantifiable, and given some assumption about the resources needed to determine the true nature of those mine-like objects, the cost of that uncertainty can be approximated. In more complex situations, where uncertainty in one or more parameters is compounded when values are used as input to mathematical models of future states or far-field conditions, understanding the impacts (i.e., the cost of uncertainty) becomes more complicated. Understanding how

A side-scan sonar unit is lowered from the high-speed vessel Joint Venture (HSV-X1) into the waters off the coast of California during Fleet Battle Experiment Juliet. Sonar was used to locate underwater mines, to enable safe navigation of amphibious forces to reach the shoreline, during exercises in support of "Millennium Challenge 2002" (Photo courtesy of the U.S. Navy).

uncertainty propagates through the various steps involved in converting data into information is an important component of the production process and of conveying the value of that information to the decisionmaker (i.e., the commanding officer).

Environmental uncertainty propagates through a chain to the commanding officer to influence his decision; moreover, there is feedback to the several links that can be used to optimize his decision process. Figure 4-3 illustrates several aspects of this. For sonar systems the useful environmental databases are typically bathymetric charts, GDEMs, DDB (Digital Data Base) at various scales, ETOPO5 by location and time of year, sound speed profiles (e.g., as generated by MODAS [Modular Ocean Data Assimilation System]), and bottom loss (BLUG [bottom loss upgrade]), plus other terms that enter the sonar equation. The coverage and resolution of these databases drive various acoustical prediction models often incorporated as tactical decision aids (SFMPL and PCIMAT) for describing how the environment modulates the acoustic propagation. Since the ocean is a very reverberant and refractive medium, the propagation can be quite complicated. One of the current perceptions is that these tools are solely limited by the fidelity and resolution of the environment.

There are, however, some realms where the propagation physics are not well modeled. If there is environmental uncertainty, there is also acoustic uncertainty, and it is not yet clear how to robustly describe how this uncertainty propagates. Certainly, there needs to be feedback from the acousticians to the environmental characterization of what is needed. Next, the acoustical output enters the signal

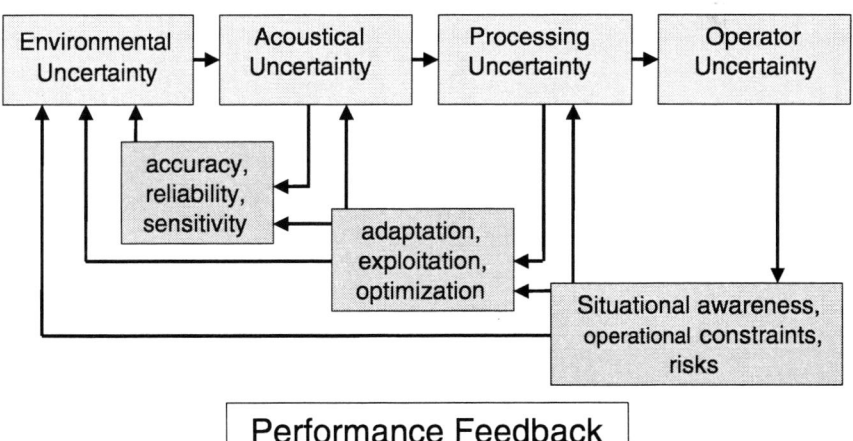

FIGURE 4-3 Propagation of uncertainty in a sonar system and the advantages that appropriate performance feedback can play in reducing its impact.

The Remote Minehunting System is an organic, off-board mine reconnaissance system that will offer carrier battle group ships an effective defense against mines by using an unmanned remote vehicle. Current plans call for the system to be first installed aboard the destroyer USS Pinckney (DDG 91) in 2004 (Photo courtesy of the U.S. Navy).

processor. While there is a good understanding of the processor, acoustical output is often so highly nonlinear with respect to errors in the acoustic models, statistical fluctuations, and system calibration errors that it is often difficult to provide a statistical prediction of its output. The so-called sonar equation is a useful guide, and receiver operating characteristics can be predicted, but they are only descriptive. There is also feedback to the rest of the components, as changing certain parameters of the signal processing can adapt and optimize the performance. Finally, the commanding officer must integrate all this uncertainty. He observes what is happening with the incoming data and must reconcile this to the prediction models. He wants to maintain his tactical advantage of position, situational awareness, and risks. If he is confident in his environmental predictions, he can exploit them to fulfill his mission objectives.

Cost of Uncertainty

In this section the goal of the METOC enterprise is assumed to be to reduce uncertainty (i.e., lack of knowledge about the nature of environmental conditions at some future time or different location) due to environmental processes, with the operational cost of that uncertainty providing guidance to optimum strategies. During tours of the committee to the various centers, members continually tried to explore the issue of expressing uncertainty in METOC. One of the most interesting insights came from a METOC officer who described his working relationship with a previous operational commander[1] in terms of a betting metaphor. With each forecast, particularly those with tactical importance, he was asked to rate his confidence in the forecast for the mission in terms of one of three categories: no bet, minimal bet, and high bet. This case of an individual (and successful) relationship between a METOC officer and his operational commander illustrates several important points. First, it recognizes the fact of uncertainty and the fact that it should affect command decisions. Second, it institutionalizes uncertainty for operational purposes in a simple set of levels (a useful example of a concept of operation or CONOPS). Third, it includes an implicit assessment of the impact of the prediction on the operation in placing the bet. That is, uncertainty in certain situations or for certain variables has no operational impact, while in other situations it is critical.

This creative solution to dealing with prediction uncertainties parallels the concepts of traditional risk analysis practiced in the business world, as discussed in Chapter 2. In risk analysis, potential adverse conditions are identified, and then estimates are made of the probability and consequences of their occurrence.

---

[1]Formerly referred to as commanders in charge or CINCs, current Department of Defense (DOD) doctrine refers to them as operational commanders with two exceptions, COMPACFLEET and COMPAC, who are still considered CINCs.

Rigid Hull Inflatable Boats stand ready for launch in the ship's well deck, in preparation for an upcoming Mine Countermeasures Exercise. Following the decommissioning of the mine countermeasure support ship USS Inchon (MCS 12), amphibious assault ships have provided transportation and support to the mine countermeasures units operating out of Naval Station Ingleside and Naval Air Station Corpus Christi (Photo courtesy of the U.S. Navy).

Consequences of Uncertainty—Relative Operating Characteristics

In the meteorology literature the consequences of prediction uncertainty have been studied using the concept of Relative Operating Characteristics (ROCs; see Mason and Graham, 1999, for a good introduction). ROCs form a basis for understanding decisions made from predictions for which confidence intervals are available (in this case through ensemble forecasts). In their discussion, Mason and Graham study the problem of issuing warnings for either drought or heavy rainfall seasons over eastern Africa, but the approach is equally applicable to decisions on whether to send the fleet to sea prior to an impending hurricane or a go/no go decision for a SEAL infiltration.

ROC analysis is based on contingency tables, matrices that compare the joint probability of the prediction and occurrence of events (see Table 4-1). For example, if wave heights were correctly predicted to be too large for a SEAL operation, the forecast would be counted as a "hit," while correct prediction of low-wave energy (no warning) is a "correct rejection." Similarly, incorrect prediction of high waves (issuing a "no go warning") is an "incorrect warning," while incorrect prediction of safe conditions is called a "miss." Each of these outcomes has a cost that can be assessed by the Navy. For example, a miss might be viewed as more expensive than an incorrect warning since the lives of SEALS could be threatened. The cost of uncertainty, then, is the cost of each prediction failure mode times the probability of its occurrence. The extension of this calculation to a range of variables, missions, and conditions is straightforward but is best illustrated by way of an example.

TABLE 4-1 Contingency Table for Risk Analysis[a]

	go	no go	Total
below threshold	correct go	incorrect hold	N
above threshold	incorrect go	correct hold	N

[a]Statistics are built up from a history of realizations of predictions and their subsequent true outcomes. The upper categories indicate cases where the forecast call was "go" or "no go" for some particular operation or action. The left-hand categories indicate the actual conditions that ensued.

Example Use of ROC to Find the Cost of Uncertainty

Consider a SEAL infiltration mission onto a sandy beach. The mission depends on a number of environmental variables. This discussion will start with an examination of one variable, wave height, first introducing uncertainty and then moving on to quantify the cost of that uncertainty. The concept will then expand to the composite cost of all variables and then the merging of costs from more than one mission.

For illustration, threshold criterion for wave height for this mission is assumed to be 1.5 m. Further assume that the predicted wave height for the time of the operation is 1.1 m. In addition, assume there is a known uncertainty to this estimate, represented by a standard deviation of 0.3 m (in the interest of brevity, discussion of possible methods for estimating confidence intervals is omitted here but is an important consideration). Thus, the actual wave height at the time of operation can be represented by a probability distribution function (see Figure 4-4). From this figure it can be seen that waves will most likely not be an impediment, and we would give a "weather go" to the operation. However, there is a 9 percent chance (marked on the graph) that the actual wave height will be greater than the threshold. Thus, if these same circumstances occurred many times, in 9 percent of these cases our prediction would be in error, and we would count these cases in a contingency table as cases of "incorrect go" (Table 4-1). Similarly, if the prediction had been for waves of 1.6 m in height, the METOC prediction would be "no go" or "hold." Most of the time this would be the correct call. However, there would again be a number of cases (37 percent) where the actual waves at the time of operation were less than the threshold. These would constitute an "incorrect hold" and would contribute to the incorrect hold quadrant of the contingency table. A full contingency table is found by computing an ensemble of example cases (see Table 4-2).

Each type of prediction error from the contingency table would have an associated cost. The cost of an incorrect go is probably judged to be high, since such a situation can lead not only to mission failure but also threat to life. On the other hand, the cost of an incorrect hold may be much lower, merely representing a missed mission. However, if the mission is critical or is a critical component of a complex interdependent mission package, the cost of an incorrect hold will rise. For example, a missed evacuation of key civilian personnel prior to a conflict may necessitate very expensive and dangerous rescue operations later on and so would be associated with a higher cost for an incorrect hold. This would not mean that the SEAL operation would be attempted in impossible conditions, but it does mean that the cost of an incorrect hold could be very high depending on the duration.[2] The impacts of missed predictions can be represented in a cost

---

[2] In reality, threshold criteria may be better viewed in terms of a green-yellow-red paradigm with a band of wave conditions where the operation becomes dangerous but still possible. Contingency tables can still be built (incorrect red, etc.) and the chance of prediction error (and associated costs) estimated. For simplicity we limit this illustration to the binary go/no go paradigm.

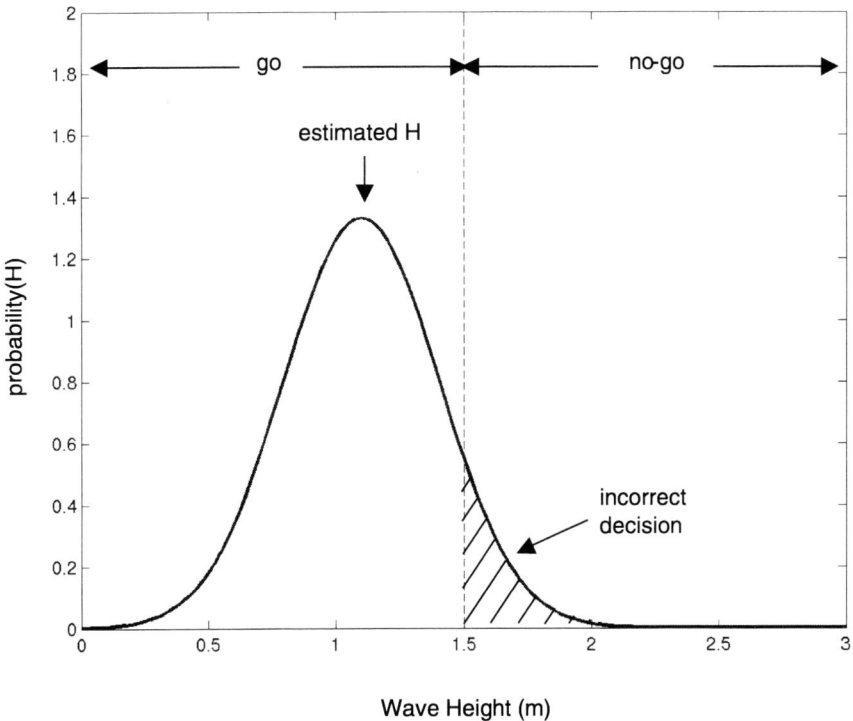

FIGURE 4-4 Example wave height prediction. The predicted value is 1.1 m for this example, with a standard deviation of 0.3 m. For this operation, the threshold for operational safety is 1.5m, so the METOC decision would be a "go". However, it can be seen that 9 percent of the time, wave heights above threshold limits would be encountered. In those conditions, the METOC decision would have been incorrect.

table (see Table 4-3) that parallels the contingency table. Values are placed in the "incorrect go" and "incorrect hold" quadrants.[3] The cost of uncertainty, then, is the sum of the likelihood of each type of failure times the cost of that failure. Correct predictions are viewed as having no cost beyond normal operations.

A question can now be asked whether the quality of wave height forecasts is sufficient or whether additional research investment is needed. Answering this question requires an assessment of the likely range of conditions that would be

---

[3]Values shown are for illustration only. The true cost of an invalid prediction would require extensive Navy input.

TABLE 4-2 Hypothetical Contingency Table Showing Accuracy of Wave Height Predictions for a Suite of Test Cases

	go	no go	Total
below threshold	56	7	63
above threshold	5	34	39
Total	61	41	102

TABLE 4-3 Contingency Cost Table[a]

	go	no go
below threshold		10
above threshold	100	

[a] There is no cost associated with successful predictions. The relative costs for "incorrect go" and "incorrect hold" depend on the impact of this variable on operations and the current risk attitude. Although it is impractical to capture all potential costs or benefits associated with even a simple military operation, some minimal assumptions must be made. This is largely done intuitively now; more formal analysis would require more formal definitions of cost and benefits.

faced by the U.S. Naval Forces for this type of mission. Thus, the entries in the contingency table will be the average overall expected conditions for some future planning life. If it is anticipated that most missions will be in enclosed seas with wave heights of less of than 0.5 m, the total likelihood of an incorrect go will be small, and the cost of the current level of uncertainty is small. Similarly, if the anticipated tactical interests lie in areas with extreme wave environments, decisionmakers will nearly always be correct in their "hold" predictions and, again, further R&D is not needed. On the other hand, if a fair portion of anticipated missions will occur in conditions that are "near" thresholds, it becomes important to be able to reduce the uncertainty of those predictions and hence reduce the rate and cost of prediction errors.

In fact, the question is not whether to invest in one variable such as wave height, but instead how to prioritize investment across the many variables and processes that affect naval forces. Continuing the case of the SEAL infiltration, it needs to be recognized that the mission is dependent on the environment in many ways. For illustration, consider the effects of water temperature and nearshore currents on the mission. One can proceed in the same way as outlined above for wave height. Uncertainty in the estimate of each variable (confidence interval) must be estimated (e.g., 3° C for water temperature). A contingency table is then created by considering the likely range of conditions to be faced in the future and thereby the likelihood of incorrect predictions (either "incorrect go" or "incorrect hold"). It is possible that the "thresholds" for different environmental variables may represent different levels of danger to the mission. Thus, there may be relative weightings of the impact of the variables, with water temperature perhaps having much less severe impact than wave height (for example). This would be represented in the cost tables, with a lower cost for a temperature "incorrect go" than for a wave height "incorrect go."

Finally, the impact of different variables on different missions can be considered. Contingency tables can be built for each mission type and environmental variable. The total cost of uncertainty, then, is the sum over the entire anticipated mission portfolio of the costs of uncertainty for each mission. For example, it may be anticipated that over the next 10 years missions will be allocated as 35 percent carrier air operations and strike warfare, 30 Naval Special Warfare, 15 percent Undersea Warfare and Anti-submarine Warfare, 15 percent Amphibious Warfare and expeditionary warfare, etc. (This distribution of activities among the various warfare missions is purely hypothetical.)0

Implementation of this concept would obviously require a moderate amount of bookkeeping and an in-depth introspection on the current capabilities of prediction systems (confidence intervals for different environmental variables) as well as the likely future mission portfolio. In addition, consideration of enemy actions, both in terms of impairing data collection or communications will need to be accounted for.[4] Much of this information does not exist. Thus, it is not necessary, or even desirable, that the operational naval METOC enterprise itself become bogged down in providing a value basis for METOC investment strategy. However, a focused effort by a qualified study group organized by the Office of Naval Research (ONR) and working closely with the Office of the Oceanographer of the Navy and the Naval Meteorology and Oceanography Command (CNMOC) could further explore this concept, laying out uncertainties and their costs for a number of the main mission problem areas.

The application of such analysis to complex systems often results in unexpected insights. The human ability to make practical and appropriate decisions in relatively simple cases involving uncertainty often results in unwarranted confidence in making decisions in more complex situations. Thus, it would be surprising if unexpected details did not emerge from a more rigorous examination of the cost of environmental uncertainty across the breadth of naval activities. For instance, it may be possible that the naval METOC enterprise now predicts environmental conditions at large spatial and temporal scales sufficiently well for most purposes that further investment would have small impact on the cost of uncertainty, compared to other uses of R&D funds. Even the process of estimation of confidence intervals will raise questions that have been ignored for too long.

Reduction of Uncertainty by Further Investment

The sections above attempt to provide a methodology for placing value on METOC knowledge and predictions by introducing the cost of uncertainty. However, the charge to the committee calls for developing a rational basis for investment in METOC R&D. Improvements in the process for providing METOC information (whether through more efficient data acquisition, incorporation of improved understanding of natural processes in forecasting tools, or other means not yet identified) will yield reduced uncertainty in predictions and hence reduced cost of uncertainty. The best investment strategy is that which provides the largest reduction in the cost of uncertainty for the smallest research investment cost. The general field of risk reduction through incorporation of improved knowledge is related to the topic of Bayesian statistics.

There are a number of methods of reducing uncertainty in a METOC prediction. The most straightforward method is to take additional measurements of the domain to be used as initial values in a model run. Such measurements can

---

[4]The uncertainties involved in any military decision are great, and many, if not most, have more to do with understanding enemy intent than understanding environmental conditions. The scope of this, however, is limited to environmental information, and evaluating the potential for actions by enemy forces to impact the ability of U.S. Naval Forces to share or leverage environmental information is more readily understandable.

include in situ sampling or remote sensing. Increased sampling may come with a cost, owing to both the cost of expendables and the associated cost of operations. It may also be either risky or impossible for denied areas of interest.

Over longer periods of time, uncertainty in predictions can be reduced through research. In general, the research can be directed toward improvement in the models (through either improved understanding of the physics or improved hardware and modeling strategies), increases in remote sensing or in situ measurement capability, or improved understanding of the optimum choice of initial value samples for model performance (model sensitivities). There are many choices on how R&D money can be invested, each with a different return value.

The cost of data acquisition varies depending on the nature of the source. The cost of collecting new data may be low in some conditions; thus, users may not pursue acquiring data previously collected by other sources. This tendency can become a liability if easy access to new data discourages the development of

An F-14 "Tomcat" fighter assigned to the "Jolly Rogers" of Fighter Squadron One Zero Three (VF-103) leads a formation comprised of F/A-18 "Hornet" strike fighters from the "Blue Blasters" of VFA-34, the "Sunliners" of VFA-81, and the "Rampagers" of VFA-83. Two Croat MiG-21 "Fishbed" fighter-interceptors flank each side of the formation. U.S. Navy aviation squadrons assigned to Carrier Air Wing Seventeen (CVW-17) have sent a detachment to Croatia in order to participate in Joint Wings 2002. Joint Wings is a multinational exercise between the U.S. and the Croat Air Force designed to practice intelligence gathering. Supporting multinational forces is an ever-increasing demand for the U.S Navy METOC enterprise (Photo courtesy of the U.S. Navy).

effective mechanisms to find and acquire data from other sources, because constantly evolving geopolitical realities can make what had previously been seen as routine data collection impractical. Thus, an evaluation of various sources of data would seem to be helpful. Furthermore, a formal analysis of the payoff for such investment can be handled through a Bayesian statistics analysis. In particular, formal decision theory can be used to examine the value of new information or improved predictions to decisionmaking.

Database Issues

As evidenced in previous chapters, METOC data needs span a broad range of data types, sources, and formats (see Box 4-1). The user community within U.S. Naval Forces spans an equally broad range of sophistication and experience in the production and use of METOC data products. To achieve the vision of a network-centric naval force as it relates to METOC issues, the supportive data systems must be fully interoperable at the machine level. This means that computers in the data system must be able to exchange data in a semantically meaningful fashion without human intervention. Furthermore, increased use by naval METOC personnel of non-Navy data sources suggests that the Navy's METOC systems would benefit if they were interoperable with non-Navy data systems (see Box 4-2).

At present, naval METOC systems are not interoperable, although elements of them may be. It is not unusual that a METOC officer requires access to several different computer displays, each addressing one data type, and the data visible on one display (system) may not be overlaid on or with data from another display (system). This is especially true for those who would like to view oceanographic data products with meteorological data products and for those interested in integrating data products generated by the Naval METOC enterprise with those generated by other DOD services or by non-DOD. The Department of the Navy (DON) does recognize the need for interoperability among its systems and is working toward this goal with the Navy Integrated Tactical Environmental Subsystem[5] (NITES, 2000) installed on 70 of the U.S. Navy's major surface ships (as well as at shore activities) is used by forecasters to process weather and ocean data from anywhere in the system. A critical element of NITES 2000 is the

---

[5]The current version of the Navy Integrated Tactical Environmental Sub-system is a modular open-architecture software subsystem that is integrated as a segment of the Navy C4I system onboard all ships and at all major Navy/Marine Corps commands and staffs, both onshore and afloat. NITES 2000 integrates derived products into command and control tactical decision aids for use with strategic and tactical computer systems on smaller ships and sites. The open-system design of NITES will provide complete interoperability with other DOD, federal, and allied command and control systems connected to the new Global Command and Control System (http://www.matthewhenson.com/usnshenson4.htm).

> **BOX 4-1**
> **Accessing METOC Data in Naval War Games**
>
> At present, METOC data and product requirements vary dramatically from user group to user group. In particular, the fleet generally requires access to near real-time data or to predicted fields, whereas war gaming requires realistic but simulated conditions. Simulated weather (both atmosphere and ocean) conditions, when used for a hypothetical conflict, are presently obtained by selecting a period in the past for which good weather observations are available for the region of interest and for which the suite of weather conditions desired for the war games occur. The observations and predictions obtained from this period are then used to simulate weather conditions for the games. The two sides in the simulated conflict have access to different-quality predictions and observations. Surprisingly, one side is provided with "perfect" predictions (i.e., if Team Blue is provided at the outset with the actual weather for the duration of the games, Team Red will be given less accurate "predictions"). In addition, regarding the need for simulated fields for actual war games, developing war game scenarios often requires access to statistical summaries or climatologies of various meteorological and oceanographic parameters (e.g., what is the probability that sea state exceeds 15 feet in the eastern North Atlantic in winter and for how long can such conditions be expected to persist?).
>
> Not surprisingly, data systems being developed by the U.S. Navy for the METOC community are focused on the needs of the fleet; the fleet is by far the largest user of such systems among U.S. Naval Forces. The result is that these data systems do not meet the broader range of war-gaming needs; specifically the systems currently under development do not provide access to a wide range of historical data, nor do they provide the capability to readily work with these data and/or with climatologies when obtained from a "non-DOD" source. As a result, generating realistic simulations for war games is tedious at best. Furthermore, the METOC community is denied an opportunity to fully evaluate the METOC data systems of the future.

Tactical Environmental Data Server (TEDS). TEDS is a "Defense Information Infrastructure (DII) Common Operating Environment (COE) compliant set (see http://www.sei.cmu.edu/str/descriptions/diicoe.html for a description of DII/COE) of database, data, and software segments that serve as the primary repository and source of Meteorology and Oceanography (METOC) data and products for NITES 2000. TEDS is composed of a METOC database and a set of Application Program Interfaces (APIs) that provide storage for and access to dynamic

BOX 4-2
Opportunities and Challenges to Deriving Environmental Observations from Imagery Collected for Intelligence Purposes

All METOC personnel are trained in the basics of weather observation. These first-hand human observations have historically formed the nucleus of the synoptic observation system that is key to forecasting. Today, satellite imagery and automated in situ weather stations provide the preponderance of observational data used by the METOC community. The electrooptical imagery collected by satellite or other means, therefore, can be considered as dislocated weather observations. Deriving a useful observation from such an image is a teachable skill that, when employed properly, reduces the electrooptical signal to a numerical parameter that can be handled just like any other measurement. Thus, the fact that these observations are derived from imagery rather than human observation or some other source cannot be determined from the derived values.

In many military situations, however, the only available imagery for a given area is that which is collected for intelligence purposes. Often, the resulting images in and of themselves are not particularly sensitive so much as the time and location or other ephemeral and associated information. The limit to deriving useful observational data from such imagery is finding adequate numbers of technical staff trained in the relevant techniques who also hold adequate clearance to work with the imagery. The utility of such capability has been explored and validated on a small scale at the Naval Pacific METOC Command-Joint Typhoon Warning Center (NPMOC-JTWC) at Pearl Harbor, Hawaii.

NPMOC-JTWC personnel with adequate clearance process imagery on a limited basis and derive valuable environmental data that are then distributed throughout the METOC community. Because all sensitive attributes have been removed, these data (which are indistinguishable from data from other sources) can be used as model input to improve forecasts over tactically important regions where other sources of information are limited. Currently, production-scale extraction of information from highly classified imagery is limited not by connectivity, technical understanding, or facility space but by the lack of resources needed to carry out security checks and other steps needed to grant existing center personnel with appropriate clearances. In other words, if adequate numbers of technically competent personnel exist, the only additional investments needed are the resources to grant them adequate clearance. Expanding the capability to other METOC centers of sea assets would involve some expanded connectivity but would largely require the commitment of resources proportional to the number of staff needed to fully integrate the approach into all relevant METOC production efforts.

METOC data (e.g., analysis/forecast grid field data, observations, textual observations and bulletins, image data, and remotely sensed data) in a heterogeneous networked environment."[6]

Despite the significant steps being taken for interoperability among naval forces and to a lesser degree between DON and other DOD services, the lack of an effort to interface these systems with non-naval, non-DOD systems is seen as limiting at present and as a potential serious omission in the not too distant future (order of 10 years; see Box 4-1).

TARGETING DATABASE DEVELOPMENT

The uncertainty model developed earlier in this chapter should be applied to the determination of what data are needed, so that a plan for what data should be collected and saved in the future can be developed. There are four potential sources of data that need to be considered in developing METOC databases: (1) environmental data currently collected but discarded after primary use; (2) data that can be collected by sensors devoted to other uses, such as intelligence gathering, surveillance, or reconnaissance; (3) data that would result from the addition of additional dedicated METOC sensors on current naval platforms; and (4) increased exploitation of nontraditional sources.

A Policy for Saving Data Currently Collected and Discarded

The DON collects a vast quantity of data with existing sensors. A subset of these data is saved for future use. The procedures for saving data do not, however, appear to be well established. In particular, data that are discarded today may prove to be important in the future.

Dual Use of Intelligence Sensors for METOC in the Littorals

Dual use refers to approaches that provide benefit for some purpose from data that were collected for a different purpose. Since the data are already in hand, there are no further costs of acquisition. Thus, dual-use techniques are generally very cost effective.

One of the best examples of the dual use of a non-METOC sensor for METOC purposes involves the Aegis radar system mounted on major combatants. The propagation and performance characteristics of this system depend on aspects of the environment, particularly atmospheric and ocean conditions that affect ducting, attenuation, and scatter of radar signals. In-dual use application this sensitivity to the environment is inverted such that radar performance can be exploited to yield estimates of characteristics of the environment of radar propagation.

---

[6]From *http://www.nrlmry.navy.mil/~lande/TEDS*.

Dual use may be particularly important in the littoral environment, a zone whose short time- and space scales of variability force frequent, high-resolution sampling for proper battlespace characterization. Since access will be denied to most areas of tactical interest, in situ sampling is dangerous and generally limited in scope. For example, Navy SEALs hold mission responsibility for bathymetry measurements in depths of less than 20 feet. Their work involves either personal observation or installation and retrieval of instruments. Both missions are clandestine and are of a type that the SEAL would rather not have to do. Thus, it is evident that the preferred METOC sampling strategy in nearshore regions should be based on remote sensing, preferably at stand-off distances.

The applicability of spaceborne data collection for describing littoral environments has been the subject of considerable recent discussion (e.g., Poulquen et al., 1997). The view that success can only be had at larger scales and that the fine-grain resolution needs for nearshore sampling make course-grained data, such as are available from most satellites, of limited use may be changing. Spatial resolutions of 1 m for hyperspectral and 8 m for SAR are now available, with daily to three-day time intervals. Some spaceborne sensors may, therefore, supply data of sufficient spatial resolution (SPOT, for example), but timely access to satellite orbit limitations, tasking conflicts, or problems with rapid dissemination is still a concern.

By contrast, these are just the scales that are needed for intelligence sensors. The U.S. Navy already owns a wide range of intelligence assets that could contribute to the METOC mission through either direct tasking or enroute supplements to normal intelligence missions. Many of these sensors are forward deployed and so are under the control of theater operational commanders. Examples of intelligence packages are TARPS or sensors on Global Hawk.

The very shallow water and surf zone regions of the littorals provide a wide range of surface signatures that can be exploited to allow estimation of variables of METOC interest (Holman et al., 1997). Methods have been developed and tested for the estimation of wave period and direction (e.g., Lippmann and Holman, 1991), the strength of nearshore currents (Chickadel et al., in review), nearshore swash (e.g., Holland et al., 1995), sand bar morphology (Lippmann and Holman, 1989), and subaerial and subaqueous beach profiles (Holman et al., 1991; Stockdon and Holman, 2000). Key to these techniques is sampling of time domain variability over a short period of dwell, usually less than two minutes (although useful but degraded results can be found for significantly shorter record lengths). This requires a sensor that can keep a region of interest in view either "in passage" or by staring.

While the above references refer to the use of optical data, similar signatures are also available from active sensors (principally radar) and from other passive bands. Each band can be, and has been, exploited for similar purposes.

Imagery and other information collected as part of intelligence-gathering efforts are generally classified as high level, introducing potential complications in transfer to and handling by METOC units. However, on aircraft carriers these activities are located adjacent to each other, and wider use of such information at the various METOC centers is limited largely by the number of technicians with adequate clearance to allow work on a production scale (Box 4-2). It seems likely that an appropriate CONOPS could be developed, as done at the Warfighting Support Center, to handle these issues. The advantages of having forward-deployed sensor capability coupled to onboard exploitation could greatly improve fleet METOC capability, particularly in a political flashpoint area, where contention for tasking of national assets can be high.

Adding Data Collection Capabilities to Navy and Marine Platforms

A simple mechanism for the improvement of naval METOC data collection capability could be the addition of small automated sensors on a range of naval platforms. Naval surface ships, submarines, and aircraft log millions of miles per year in all of the world's oceans. Naval aircraft fly up and down the atmospheric air column thousands of time per day. Ships sail routinely in areas where surveys are either very old or nonexistent and bottom obstruction mapping has not been attempted for decades. Submarines are diving through the water column through-

Although modern technology has greatly enhanced efforts to provide timely and accurate forecasts, lookouts on surface vessels still provide valuable in situ information on sea state and current weather conditions (Photo courtesy of the U.S. Navy).

out the world and in most cases are not recording or saving valuable environmental data.

Few U.S. Navy or Marine Corps operational units have any automated environmental sensors or the equipment required for processing and storing data. There are several barriers to deployment of these types of systems. Initially, sensing systems tended to be big, expensive, power hungry, and fragile. Accurate positioning and precise time were not readily available to automatically register data in time and space. Finally, there was the question of who would pay for development and deployment of such equipment. Generally speaking, the various warfare communities—surface, submarine, and aviation—were not interested in budgeting for items that would not directly contribute to warfighting capabilities.

Today, lightweight, rugged, low-cost, low-power environmental sensors are available off the shelf. The Global Positioning System (GPS) now provides accurate locating data and precise time that can easily be integrated into computer-based systems. Personal computers are not expensive and can quickly register new data and store them on internal hard drives that can handle up to 100 gigabytes. Attached storage units are now in the terabyte range. In practical terms, a year's worth of METOC observations and bathymetric images could be stored on one hard drive and easily be backed up on CD-R or DVD-R drives. This information can serve as a valuable resource for intelligence, navigation, climatology, and scientific research.

The most useful data for surface ships would be air and sea surface temperature, humidity, water column bathymetry, and bottom swath bathymetry. Aircraft should be extracting temperature, humidity, lapse rates, and altitude. Submarines could easily extract conductivity, temperature, and depth, the basic building blocks of oceanographic research. This information could be displayed locally in graphic format for tactical use and passed by datalink for strategic modeling and research onshore.

Other Nonstandard Data Sources

One of the most important assets to U.S. Naval Forces is personnel. METOC officers gain a great deal of knowledge related to the meteorology and oceanography of the regions they visit, as do the enlisted personnel who support them. Some of this experience is captured and saved in cruise reports generated by METOC officers. There is also potentially useful information exchanged by METOC officers in online chat sessions[7] across SIPRNET (referred to as IRC

---

[7]It is important to note here that these communications, though referred to by METOC personnel as "chat" sessions, have very little in common with the chat rooms readers may be familiar with. Senior naval personnel frequently participate in or follow these discussions and, though the messages are loosely structured, the fraction of irrelevant or casual conversation is extremely low and the message content is consistent with other sensitive traffic carried on SIPRNET.

Chat, these interactions have become an extremely popular and effective mechanism for forward-deployed METOC officers to interact with colleagues stateside or on other naval platforms. Since these sessions are electronic, the information all exists in a form that is easily saved. At present, though, this information is either not being saved or is not organized for easy access. This is because systems do not exist today that can effectively mine such databases. There is, however, a great deal of research being undertaken in the mining of textual databases with significant progress being made, and it is not unreasonable to think that in the not too distant future (less than 10 years) sufficient progress will have been made for the naval METOC community to make more effective use of these data.

Non-DOD Data Repositories

As discussed in Chapter 3, data systems under development for the METOC community are focused on access to relevant data and products generated by DON efforts and/or passed through naval (and in some cases other DOD) data centers. There is little to no consideration in the development of these systems for direct access to data that are generated and held outside of DOD despite the use of such data sources via standard Web browsers by METOC personnel in recent naval operations. The trend toward increased development of and open access to real-time data and data products in the commercial and non-DOD research sectors can be attributed to two factors: (1) substantial improvement in computational power available at costs that are affordable to the academic researcher or the commercial data provider, and (2) the development of instrumentation that provides real-time access to data obtained at remote sites. Virtually every oceanographic and meteorological research institution is involved in one way or another with the development of smart sensors that relay their data back to the home institution in near real time. At the same time there is increasing interest in incorporating these data into data assimilation experiments and regional predictions. There is also increased use by the research community in real-time feeds from major satellite (and other) data systems in data product generation using sophisticated retrieval algorithms and/or advanced assimilation techniques. These state-of-the-art products are often the best available at the time for the region of interest. Good examples of this are the TRMM (Tropical Rainfall Mapping Mission), sea surface temperature, and QuikScat and SSM/I (Special Sensor Microwave/Imager) wind data products available in near real time from Remote Sensing Systems, a private research company (http://www.ssmi.com).

At the same time that non-DOD near real-time data sources are coming online, there is growing interest in providing seamless access to such sites. For example, the National Virtual Ocean Data System (http://nvods.org) currently provides access in a consistent form to more than 300 datasets stored in a variety of formats from approximately 30 sites in the United States, ranging from government facilities to private companies to academic research institutions. Recently several sites have also been established abroad (France, Great Britain, Australia, and Korea). As more clients are developed for this system (there are currently eight application packages from which data served via the system can be accessed), there is more interest in adding sites. At present (May 2002), new sites are being added in the United States and abroad every two to three weeks.

Given current trends in data availability and data product generation, together with the development of data systems designed to provide seamless access to data in a distributed heterogeneous environment taking place outside of the DON, it seems clear that in the not too distant future the non-DOD research and commercial sectors both in the United States and abroad could become a significant source of useful METOC data for naval forces. Surprisingly, however, there does not appear to be any effort in the development of DON systems to provide for interoperability with systems in the non-DOD sector.

RAPID ENVIRONMENTAL ASSESSMENT

As the pace of warfare increases during the 21st century, there is an emerging need for Rapid Environmental Assessment as an aid to warfighters. Whereas in the past the decision cycle for warfare operations might have been weeks to days, in the 21st century the decision cycle for many naval operations is compressed into hours and sometimes minutes (see Figure 4-5).

Rapid Environmental Assessment Versus Optimized Environmental Characterization

Discussions throughout this study often return to the fundamental utility of environmental data and use of such data by warfighters as an exploitable component of modern warfighting doctrine. In this regard, two modes of environmental information acquisition and use were identified: Rapid Environmental Assessment (REA) and Optimized Environmental Characterization (OEC). Both are explored briefly below with the objective of providing readers with a working definition of each.

Assessing the environment within the 4-D battlespace is a major task facing the naval METOC community. In particular, as the pace and scale of warfare operations increase and become more spatially fragmented, providing accurate environmental information in a timely manner so that warfighters can meaningfully exploit their environment becomes a daunting task. REA refers to environmental information gathered through networked battlespace sensor arrays and timely processing of environmental information into meaningful data products that can be disseminated to fleet assets. As such, REA is fundamentally a method for providing warfighters with synoptic views of the battlespace. Such views are useful in directing fleet or other warfighting assets during the prosecution of specific warfare operations. Ideally, these views of the battlespace are updated

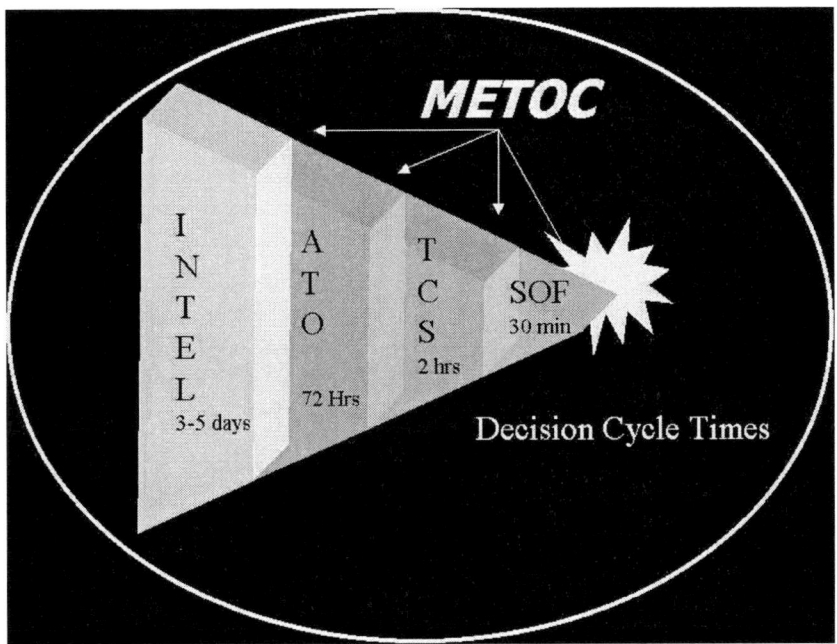

FIGURE 4-5 Decision cycle times for typical naval operations requiring environmental information.

frequently (i.e., every few minutes) to provide military commanders with sequential synoptic views of the evolving battlespace environment. The Naval Fires Network (NFN) is an example of a weapons system developed to utilize REA concepts.

OEC differs somewhat from REA in that a goal of OEC is to utilize REA information to parameterize various models and develop predictions that forecast the evolution of the battlespace. As such, OEC is fundamentally a method for providing warfighters with a predictive view of the battlespace. A critical component of a successful OEC system is the ability to ingest high-priority REA data (in other words, optimize the process) in order to compute predictions about the future environmental state of the battlespace and receive continuous updates of battlespace conditions from REA sensors in order to dynamically update battlespace forecasts. Such forecasts not only are useful for planning scenarios but may also ultimately find use during naval combat operations to dynamically modify warfighting strategy in order to gain continuous tactical advantages or minimize threats to fleet assets. For example, an array of REA sensors in a battlespace may transmit information regarding wind speed, direction, and vertical wind structure in the atmosphere. These data might be merged with high-resolution terrain models in an OEC system designed to predict and forecast dispersion of biological, chemical, or radiological agents at a pace set by the operational tempo. Whereas the synoptic view of a dispersing plume provided by REA may aid naval commanders in understanding conditions occurring within the battlespace, OEC model output would help commanders make decisions regarding where to move assets in order to minimize the effects of a plume from weapons of mass destruction on warfighters. Furthermore, OEC output could be used to predict the rate of dispersion of such a plume, providing commanders with information about when it might be safe to move assets back into the affected area.

It is clear from the brief discussion above that REA and OEC, though somewhat distinct concepts, are also closely related and that successful application of one depends on the relative state of development of the other. As such, REA and OEC should be viewed as separate but complementary elements of a more thorough naval environmental information system.

REA represents the need for theater-wide environmental information gathering through sensor arrays and timely processing of environmental information

A Seabee assigned to Beach Master Unit Two based at Little Creek, Va., directs a Landing Craft Air Cushion onto Onslow Beach at Camp LeJeune, N.C. Understanding the environmental thresholds for a variety of naval platforms will be key to how to enhance current efforts to provide timely and valuable information to U.S. Naval Forces (Photo courtesy of the U.S. Navy).

TABLE 4-4 Naval Mission Areas Where REA May Be a Critical Component

Mission Area	Acronym
Anti-Air Warfare	AAW
Amphibious Warfare	AMW
Anti-Surface Warfare/Over-the-Horizon Targeting	ASU/OTHT
Command/Control/Communications/Computers, Intelligence, Surveillance and Reconnaissance	C4ISR
Operations Other Than War	OOTW
Naval Special Warfare	NSW
Strategic Deterrence and Weapons of Mass Destruction	STRAT/WMD
Strike Warfare	STRIKE
Wargames and Training Issues	WGT

into meaningful data products that can be disseminated to fleet assets. Some naval warfare operations (e.g., time-critical strike and special warfare) have evolved a need for environmental information that is both comprehensive and rapidly updated to ensure success.

Development of all-weather precision munitions (e.g., GPS-guided weapons) minimizes the need for rapidly updated environmental data since weapon delivery depends on accurate target locations rather than specific environmental conditions (such as an ability to see the target).[8] However, some warfare operations have become more dependent on timely environmental information (e.g., time-critical strike, special warfare operations, ship self-defense, weapons of mass destruction, etc.).

Further, reducing the exposure time of naval forces during combat operations is among the most effective strategies to minimize combat losses. As such, reliable and up-to-date environmental information is critical to attaining mission objectives. Examples of naval warfare operations where REA is an increasingly critical component are listed in Table 4-4.

In many of the operations listed in Table 4-4, REA capabilities to provide near real-time assessment of environmental conditions throughout the 4-D battlespace would greatly enhance probabilities for successful execution of mission tasks. For example, emerging capabilities to engage in time-critical strike warfare (TCS) against mobile or elusive targets would greatly benefit from the ability to rapidly and accurately assess and evaluate environmental conditions such as atmospheric radio frequency propagation characteristics, visibility over target

---

[8]Of course, relying on GPS-guided munitions reduces the need to accurately predict visibility over the target. It also introduces a need to better understand space weather, which can introduce uncertainty (potential error) in GPS because of ionospheric scintillation. The same can be said for space weather affecting communications and surveillance capabilities.

areas, atmospheric refractive properties, surface to high-altitude winds (both direction and velocity), relative humidity, and so forth. Rapid assessment and ingestion/processing of near real-time environmental data would have similar benefits to dispersion models used to determine "red zones" of weapons of mass destruction.

To make REA a reality across multiple mission areas, there is a broad spectrum of R&D issues that need to be addressed by the METOC community and these issues need to be integrated into the overall mission of naval METOC. The principal issues related to developing REA capabilities by naval METOC are related to (1) environment assessment, (2) sensor optimization, and (3) determining customer needs. Each of these topics will be addressed briefly below to provide a sense of the present state of development and areas where significant advances in R&D might be achieved in the future.

Marines with Company D, Light Armored Reconnaissance, Battalion Landing Team 3/1, 11th Marine Expeditionary Unit (MEU) (Special Operations Capable), focus down range during recent live-fire training. With the aid of their Light Armored Vehicle Two Five (LAV-25), the Marines can undertake a number of missions, to include facilitating reconnaissance, artillery direction, and hit-and-run missions. Each LAV-25 is equipped with a 25-mm chain gun and two M-240E1 machine guns, enabling the Marine gunners to accurately fire on targets while moving at speeds of up to 10 mph due to the vehicle's stabilization system. Providing timely and valuable environmental information to such forward-deployed and highly mobile units presents a challenge to the U.S. Navy and Marine METOC communities (Photo courtesy of the U.S. Marine Corps).

ENVIRONMENTAL ASSESSMENT

Assessing the environment of the 4-D battlespace remains among the most vexing tasks facing the naval METOC community. In particular, as the pace and scale of warfare operations increase and become more spatially fragmented, providing accurate environmental information in a timely manner so that warfighters can meaningfully exploit their environment becomes a daunting task. Relevant issues in the development of REA capabilities are questions regarding:

1. determination of appropriate environmental parameters for different mission areas,
2. the granularity (resolution and sampling frequency) of gathered environmental data,
3. sensor optimization (types of sensors and optimized sensor arrays) for obtaining environmental information at the appropriate granularity,
4. quality control and quality assurance of rapidly acquired environmental data,
5. processing/analysis of data,
6. development and dissemination of relevant derived products, and
7. personnel training in the use and significance of environmental data products.

REA Parameters

Defining appropriate REA parameters requires close integration among the naval METOC community and it customers, the warfighters, as well as sensor developers. For each mission area, a decision system identifying those required parameters, desired parameters, and relatively unimportant parameters could be available (the General Requirements Data Base is a start toward a more robust parameter identification system).

REA Granularity and Sampling Frequency

REA data acquisition and assimilation needs for improved modeling scales and refresh rates need to be identified (see Tables 4-5 and 4-6). These needs are applicable to aerosol modeling efforts, high-resolution tide/surf/wave models, atmospheric dispersion models, etc. In addition, data needs for data-sparse regions of the earth will need to be identified and strategies to populate these regions with sensor arrays developed.

Additional data are necessary in order to enhance initialization of models at all scales and are particularly important as the forecast frequency becomes shorter (from 12 hours to 6 hours or less). Denser sensor arrays will also yield improved tropical storm intensity forecasts and improved storm track prediction.

TABLE 4-5 Present REA Granularity and Production Cycle

Atmosphere/ocean modeling scales
Global: 81 km (FNMOC)
Regional: 27, 9, 3 km (FNMOC)
"On scene": 27, 9, 3 km (FNMOC, regional centers)
Nowcast: 3 km and less

Atmosphere/ocean timescales
Global: 24 to 144 hours
Regional: 12 to 48 hours (72 hours possible)
"On scene": 12 to 48 hours (72 hours possible)
Nowcast (6.2 R&D): 0 to 6 hours

TABLE 4-6 Future REA Granularity and Production Cycle

Atmosphere/ocean modeling scales
Global: 27 km (FNMOC)—coupled
Regional: 3, 1, 0.3 km (FNMOC)—coupled
"On scene": 3, 1, 0.3 km (FNMOC, regional centers)
Nowcast: 3 km and less

Atmosphere/ocean timescales
Global: 24 to 240 hours (ensembles)
Regional: 72 to 96 hours
"On scene": 72 to 96 hours (72 hours possible)
Nowcast (6.2 R&D): 0 to 6 hours

SENSOR AND SENSOR ARRAY OPTIMIZATION

The need for environmental data on short timescales throughout the 4-D battlespace may drive the development of new sensors or multisensor packages or new ways to deploy sensors (e.g., unmanned airborne vehicles [UAVs] and UUVs). Obviously, the necessity for acquiring environmental data will be weighed against other options (such as deployment of weapons systems on UAVs and UUVs), and a decision system for selecting appropriate sensors will need to be developed. Optimization of sensors and sensor arrays will again require close coordination between METOC personnel and warfighters to ensure that appropriate data are collected and transformed into meaningful METOC products.

Quality Control and Quality Assurance of REA Data

Acquiring large quantities of environmental data will require development of new algorithms to assess and assure data quality. While this process might be

Timely weather forecasts play a critical role in the safe operation of aircraft at sea (Photo courtesy of the U.S. Navy).

most efficiently developed utilizing an automated system of some kind, it should be recognized that in some instances data outliers may represent significant perturbations in the environment that might be of interest to warfighters because those perturbations could significantly impact a mission. Thus, mechanisms for recognizing these perturbations and including or excluding them from model runs should be developed. At the very least, some system for determining whether anomalous environmental readings represent real environmental anomalies or sensor malfunctions should be developed. Quality control and quality assurance schemes will also aid in verification and validation of high-frequency forecast models.

Processing and Analysis of Data

Present capabilities in the naval METOC community to provide REA are evolving, and several successful trial programs are being evaluated (e.g., Distributed Atmospheric Mesoscale Prediction System, or DAMPS, and NFN). Each of these REA test systems relies on through-the-sensor data gathering, assimilation, processing, and dissemination. In each case, sophisticated METOC computer models are forward deployed such that on-scene modeling capabilities provide warfighters with enhanced environmental information related to the immediate area of operations. Each of these systems has reach-back capabilities or can serve as the primary data-gathering/processing node and is capable of providing a common operation picture through network dissemination.

Development of Relevant Data Products

The METOC community needs to receive frequent evaluations of its products from the end-user community at sea to determine the relevance of data products forwarded to the fleet. During wartime, METOC products should have a high degree of customizability such that different actors in the fleet might be able to extract environmental information most relevant to their immediate warfighting needs. For instance, naval aviators involved in close air support and amphibious assault groups have a need for access to near-surface wind data, but each actor needs to be able to interpret wind data differently. Aviators need to know how wind speed and direction might affect weapons performance, whereas amphibious assault personnel need to know the effect of wind velocity and direction on sea state. In each instance the available data need to be presented as relevant products.

Personnel Training

As the sophistication and complexity of forward-deployed REA sensors, sensor arrays, processing/analytical capabilities, and forecast products increase, there will be an increasing need for highly trained analysts to accompany the fleet in order to provide interpretive expertise. The naval METOC community should attempt to identify or develop personnel for these roles and implement career reward systems for those individuals who may serve in these billets. A recurring theme among a number of the sites visited by the committee was the lack of a career path for navy aerographers and METOC specialists. In view of the evolving importance of METOC (especially forward-deployed REA capabilities), the Office of the Oceanographer of the Navy in particular should initiate a plan to enable motivated personnel to pursue this career path.

Statistical and Other Approaches to Decisionmaking in the Face of Uncertainty

Many fields of endeavor face the consequences of decisionmaking in the presence of uncertainty (for a good introduction, see Berger, 1985). One means of providing quantitative support to decisionmaking in the presence of uncertainty is through the use of Bayesian statistical analysis (Bayesian statistical analy-

Traditional means of atmospheric sampling are both costly and manpower intensive. New capabilities in remote sensing provide for more timely and efficient synoptic measurements (Photo courtesy of the U.S. Navy).

sis is a distinct field from Bayesian decision theory). This section discusses the problem of building models of complex systems having many parameters that are unknown or only partially known and the use of Bayesian methodology. There are numerous papers and books related to the application of Bayesian analyses that have some implications for using Bayesian statistical approaches to uncertainty analysis (e.g., Berger, 1985; Gelman et al., 1995; Punt and Hilborn, 1997). This discussion is intended simply to suggest that such techniques may have value in helping set priorities for data collection under a variety of conditions. Thus, while aspects of these approaches are discussed here, the reader is encouraged to more fully explore these through the rich literature that exists.

There are three major elements in the Bayesian approach to statistics that should be indicated clearly: (1) likelihood of describing the observed data, (2) quantification of prior beliefs about a parameter in the form of a probability distribution and incorporation of these beliefs into the analysis, and (3) inferences about parameters and other unobserved quantities of interest based exclusively on the probability of those quantities given the observed data and prior probability distributions.

In a fully Bayesian model, unknown parameters for a system are replaced by known distributions for those parameters observed previously, usually called *priors*. If there is more than one parameter, each individual distribution, as well as the joint probability distributions, must be described.

A distinction must be made between Bayesian models, which assign distributions to the parameters, and Bayesian methods, which provide point estimates and intervals based on the Bayesian model. The properties of the methods can be assessed from the perspective of the Bayesian model or from the frequentist[9] perspective. Historically, the "true" Bayesian analyst relied heavily on the use of priors. However, the modern Bayesian has evolved a much more pragmatic view. If parameters can be assigned reasonable priors based on scientific knowledge, these are used (Kass and Wasserman, 1996). Otherwise, "noninformative" or "reference" priors are used.[10] These priors are, in effect, designed to give resulting methods properties that are nearly identical to those of the standard frequentist methods. Thus, the Bayesian model and methodology can simply be routes that lead to good statistical procedures, generally ones with nearly optimal frequentist properties. In fact, Bayesian methods can work well from a frequentist perspective as long as the priors are reasonably vague about the true state of nature. In addition to providing point estimates with frequentist optimality properties, the posterior intervals for those parameter estimates are, in large datasets, very close to confidence intervals. Part of the modern Bayesian tool kit involves assessing the sensitivity of the conclusions to the priors chosen, to ensure that the exact form of the priors did not have a significant effect in the analysis.

There are two general classes of Bayesian methods. Both are based on the posterior density, which describes the conditional probabilities of the parameters

---

[9]Frequentist statistical theory measures the quality of an estimator based on repeated sampling with a fixed nonrandom set of parameters. Bayesian statistical theory measures the quality of an estimator based on repeated sampling in which the parameters also vary according to the prior distributions. Most beginning statistics courses focus on frequentist methods such as the *t* test and analysis of variance.

[10]Such priors are sometimes "improper" in that the specified prior density is not a true density because it does not integrate to 1. A prior distribution is proper if it integrates to 1.

given the observed data. This is, in effect, a modified version of the models' prior distribution, where the modification updates the prior based on new information provided by the data. In one form of methodology, this posterior distribution is maximized over all parameters to obtain "maximum a posteriori" estimators. It has the same potential problem as maximum likelihood in that it may require maximization of a high-dimension function that has multiple local maxima. The second class of methods generates point estimators for the parameters by finding their expectations under the posterior density. In this class the problem of high-dimension maximization is replaced with the problem of high-dimension integration.

Over the past 10 years, Bayesian approaches have incorporated improved computational methods. Formerly, the process of averaging over the posterior distribution was carried out by traditional methods of numerical integration, which became dramatically more difficult as the number of different parameters in the model increased. In the modern approach the necessary mean values are calculated by simulation using a variety of computational devices related more to statistics than traditional numerical methods. Although this can greatly increase the efficiency of multiparameter calculations, the model priors must be specified with structures that make the simulation approach feasible.

Although they are not dealt with extensively here, a number of classes of models and methods have an intermediate character. For example, there are "empirical Bayes" methods in which some parameters are viewed as arising from a distribution that is not completely known but rather known up to several parameters. There are also "penalized likelihood methods" in which the likelihood is maximized after addition of a term that avoids undesirable solutions by assigning large penalty values to unfeasible parameter values. The net effect is much like having a prior that assigns greater weight to more reasonable solutions and then maximizes the resulting posterior. Another methodology used to handle many nuisance parameters is the "integrated" likelihood in which priors are assigned to some of the parameters to integrate them out while the others are treated as unknown. This provides a natural hybrid modeling method that could have fishery applications.

Limitations of Applying Formal Decision Theory

The case for Bayesian methods presented above must be tempered with some limitations of any formal approach to decision theory. Bayesian decision and risk-benefit analysis needs an assignment of the a priori and transition probabilities as development of a cost matrix. This has led to many historical and philosphical discussions about the foundations of this approach. How does one assign a cost matrix? How does one assign a cost to casualties? Often political risks defy a cost assignment that is needed when calculating the posterior probability of mission success or failure.

Similarly, the probabilities must be assigned. Does one estimate prioris from a database? If so, how much variability is there in the estimate? What is an appropriate assignment in a changing environment? Extremals (i.e., low-probability events) are especially vexsome because they happen so rarely. Transition probabilities are also problematic. The outcome of an observation can often depend on the system used or a tactical decision aid whose performance is questionable given an environmental database. Again, how does one quantify these probabilities and incorporate the variability into the analysis?

The important point here is that the inputs needed by a Bayesian-like approach are often not readily available or precisely defined without error. Estimating and/or assigning these are difficult problems in their own right. There needs to be an investment and infrastructure to quantify the needed probabilities and costs for this approach. This cannot be done on intuition or subjectively for this renders the results equally subjective. Certainly, one can give conditional outputs and one of the important aspects of the Bayesian approach is quantitative consideration of the inputs.

Whether setting priorities based on some cost-benefit analysis or simply determining which actions are the most pressing involves placing some value on the outcome. Such formalized efforts require considerable effort to establish a cost function associated with any given action (or even lack of action). In addition, understanding the transfer function (how costs or benefits are accrued) requires detailed understanding of how various processes interact. In a complex system, such as the battlespace, simplifying assumptions can reduce the numerical complexity but also introduce the possibility of reducing the problem to unrealistic levels. At some point, the resulting numerical model simply becomes invalid. Overcoming these barriers will require a consistent and committed effort, one that is likely best undertaken by ONR, working closely with the METOC community and warfighters. Such an effort, however, would likely bring greater focus and awareness of exactly how important specific environmental information is or can be.

In addition, Bayesian approaches present several important issues of which the user should be aware, including the following:

1. Specification of priors in a Bayesian model is an emerging art. To do a fully Bayesian analysis in a complex setting, with no reasonable prior distributions available from scientific information, requires a careful construction whose effect on the final analysis is not clear without sensitivity analysis. If some priors are available but not for all parameters, a hybrid methodology, which does not yet fully exist, is preferred.

2. In a complex model it is known by direct calculation that the Bayesian posterior mean may not be consistent in repeated sampling. However, little information is available to identify circumstances that could lead to bad estimators.

3. Although modern computer methods have revolutionized the use of Bayesian methods, they have created some new problems. One important practical issue for Monte Carlo-Markov Chain methods is construction of the stopping rule for the simulation. This problem has yet to receive a fully satisfactory solution for multimodal distributions. Additionally, if one constructs a Bayesian model with noninformative priors, it is possible that even though there is no Bayesian solution in the sense that a posterior density distribution does not exist, the computer will still generate what appears to be a valid posterior distribution (Hobert and Casella, 1996).

4. There is a relative paucity of techniques and methods both for diagnostics of model fit, which is usually done with residual diagnostics or goodness-of-fit tests, and for making methods more robust to deviations from the model. Formally speaking, a Bayesian model is a closed system of undeniable truth, lacking an exterior viewpoint to make a rational model assessment or to construct estimators that are robust to the model-building process. To do so and retain the Bayesian structure requires constructing an even more complex Bayesian model that includes all reasonable alternatives to the model in question and then assessing the posterior probability of the original model in this setting.

Despite progress in the numerical integration of posterior densities, models with large numbers of parameters can be difficult to integrate. Two of the most commonly used methods—ampling importance resampling and Monte Carlo-Markov Chain—have difficulty with multimodalor, highly nonelliptical, density surfaces.

Other Approaches

There is an ongoing effort to address uncertainty by a variety of means. One approach that may deserve additional investigation in its application to environmental uncertainty is "fuzzy arithmetic." Fuzzy numbers have been described by Ferson and Kuhn (1994) as a generalization of intervals that can serve as representations of values whose magnitudes are known with uncertainty. Fuzzy numbers can be thought of as a stack of intervals, each at a different level of presumption (alpha), which ranges from 0 to 1. The range of values is narrowest at alpha level I, which corresponds to the optimistic assumption about the breadth of uncertainty.

Using fuzzy arithmetic software minimizes the possibility of computational mistakes in complex calculations. In summary, fuzzy arithmetic is possibly an even more effective method than interval analysis for accommodating subjective uncertainty; its utility should be examined more carefully for use in addressing uncertainty in environmental information for naval use.

SUMMARY

ROC is discussed above as a possible method for placing value on METOC information, and decision theory is examined for its potential value for quantifying optimum investment strategies. The committee was not familiar with examples of these approaches having been applied to the type of environmental problems faced by the U.S. Navy and Marine Corps—with one notable exception. ONR, which supported much pioneering work on decision theory, recently made capturing uncertainty in acoustic propagation the focus of a department research initiative (DRI). This DRI, entitled "Capturing Uncertainty in the Common Tactical/Environmental Picture," is scheduled to continue through 2004 and is intended to "to characterize, quantify, and transfer uncertainty in the ocean environment to calculations of acoustic fields and to the subsequent use of acoustic fields in performance prediction, in estimating and displaying the state of targets, and in other Navy relevant applications." This initiative, though early in its history, appears to offer some opportunity to address a significant tactical problem in undersea warfare (see Chapter 2 for more discussion). ONR should expand this initiative to cover other environmental problems and processes in order to more directly support operational challenges facing the fleet and Marine Corps and the METOC community supporting them.

In the absence of more specific knowledge of the ultimate payoff from such focused ONR activities, this introductory discussion should serve to raise a number of important points concerning METOC within the U.S. Navy and Marine Corps. Foremost is the need to recognize uncertainty in METOC prediction as fundamental to the value system for METOC investment. In the absence of some method to quantify the cost of METOC uncertainty—that is, the risk due to METOC prediction failure and the likelihood and cost of that failure—no objective basis can be found for R&D investment strategies or data acquisition.

As discussed previously, uncertainty can be reduced in many instances by the acquisition of additional data. However, the cost of such acquisition may exceed the benefit derived from its use. In some instances the stochastic nature of some processes means that some uncertainty will remain regardless of the number of observations made (objective uncertainty). In other cases increased understanding of the system, whether through acquisition of additional data or better understanding of the process involved, may be impractical due to lack of assets or time. In other cases, it may be practical to collect additional data, but the need for more accurate predictions of future states is low (i.e., naval operations are not sensitive to the process involved), making additional data collection unnecessary. Understanding the nature of the uncertainty, and its potential cost, associated with any environmental parameter of interest is a key step in improving the quality of environmental information provided to the fleet and Marine Corps. War gaming, SeaTesting, and other approaches to developing naval doctrine are important opportunities to more fully evaluate the importance of reliable and accurate environmental information and the impact of uncertainty or efforts to reduce the cost of that uncertainty. The Oceanographer of the Navy and the Chief of Naval Research should make every effort to ensure that environmental conditions and predictions of those conditions are realistically depicted in simulations and that their effects are faithfully and accurately captured so that the fleet and Marine Corps, as well as the METOC community, can more fully evaluate the importance of METOC information in naval operations.

Once a decision for additional information has been made, it seems sensible to assume that the cost of acquiring those data is a function of their availability. It appears that too often data collected previously, whether for other purposes or not, are not made use of by the METOC community. Thus, the benefit of collecting those data is not fully utilized. Several opportunities for less expensive data acquisition bear fuller evaluation. Because data for denied areas are often the most valuable, especially for REA, particular emphasis should be placed on making fuller use of data collected in these areas. Typically, the best source of such data is the electronic intelligence community, as it possesses the most capable assets. Many data collected in these areas are classified simply because their recovery involves acknowledgement of sensitive ground assets (e.g., it appears that information from UAVs collected for the Defense Threat Reduction Agency [DTRA] may be held for an extended period of time before it is released to METOC archives, so as not to release information about the existence of information-gathering assets).

Current sensors provide data not only in optical bands but also from active sensors (radar) and in other passive bands. Each provides a new window into the environment. Exploitation techniques are already available in the literature and could easily be forward deployed. It is strongly recommended that existing intelligence-gathering sensors be examined for potential exploitation as dual-use METOC sensors. Particular emphasis should be placed on airborne sensor packages, since these are forward-deployed theater assets that are already under the control of operational commanders. Data collection could be either specifically tasked or en route to other missions, while analysis could be easily handled onboard.

5

Information Flow: Leveraging Network-Centric Concepts

> This chapter points out:
>
> - the significant characteristics of network-centric warfare;
> - how the naval meterological and oceanographic (METOC) enterprise may be impacted by the transformation taking place within the Department of Defense (DOD), specifically with regard to network-centric warfare;
> - the benefits that should be accrued by both the METOC enterprise and U.S. Naval Forces (U.S. Navy and Marine Corps) as each more fully embraces network-centric principles; and
> - what steps should be taken to fully capitalize on this opportunity for change.

The implications of Joint Vision 2020 (Department of Defense, 2000), future naval operational concepts, and the spread of advanced technologies and commercial information systems worldwide make it inevitable that joint forces, particularly forward-deployed naval forces, must move toward network-centric operations. In its report, *Network-Centric Naval Forces: A Transition Strategy for Enhancing Operational Capabilities,* the National Research Council (2000a), defined such operations as follows: "Network-centric operations are military operations that exploit state-of-the-art information and networking technology to integrate widely dispersed human decisionmakers, situational and targeting sensors, and forces and weapons into a highly adaptive comprehensive system to achieve unprecedented mission effectiveness."

As also discussed in *Network-Centric Naval Forces*, forward deployment of naval forces (see Box 5-1) that may be widely dispersed geographically, the use of fire and forces massed rapidly from great distances at decisive locations and times, and the dispersed highly mobile operations of Marine Corps units are examples of future tasks that will place significant demands on networked forces and information systems. Future naval forces must be supported by a shared and consolidated picture of the situation, distributed collaborative planning, and battlespace control capabilities. In addition, the forces must be capable of coordinating and massing for land attacks and of employing multisensor networking and targeting for undersea warfare and missile defense. This capability for enhancing coordinated and massed attacks emphasizes the important role network-centric operations will have in improving adaptability (i.e., the ability to configure the force in ways that were not anticipated in advance).

Network-Centric Naval Forces makes a compelling case that the trend toward network-centric operations is inevitable:

> One reason is the pull of the opportunity: The anticipated effectiveness of joint, networked forces is compelling. A second is the push of necessity: Threats are becoming more diverse, subtle, and capable. If they are to be discerned, fathomed, and effectively countered in timely fashion, increasingly complex information gathering and exploitation will be required. Also the diversity and geographic spread of potential threats and operations, many of which will occur simultaneously or nearly so, demand that forces of any size be used to their maximum effectiveness and efficiency. Another reason derives from the relentless advance of U.S. and foreign technology in both the civilian and military spheres: There will be no other way for U.S. forces to develop. Only a force that is attuned to and capable of harnessing the power of the information technology that drives modern society will be able to operate effectively to protect that society.

The comprehensive nature of network-centric naval operations envisioned in *Network-Centric Naval Forces* will lead to a new system structure, as shown in Figure 5-1. A structure that deemphasizes the one-way stovepipe flow of information that characterizes the current METOC system (see Chapter 2). When an approach that emphasizes lateral and parallel information flow is applied to information gathering and command dissemination, it may manifest itself in an architecture similar to that depicted in Figure 5-2.

INTERACTIVE INFORMATION FLOW

The previous discussion of the impact of network-centric concepts on naval operations implies that METOC operations within the U.S. Navy and Marine Corps will be forced to adapt in many unforeseen ways. Today, every military unit is (or can be) equipped with a unique Internet protocol (IP) address, thus enabling innovative distributed sensor nets. It is likely that there are many sensors

> **BOX 5-1**
> **Understanding the Role of U.S. Naval Forces**
>
> *Technology for the United States Navy and Marine Corps, 2000-2035* (National Research Council, 1997) projected that future naval forces would continue to be required to perform tasks such as the following (Vol.#1, *Overview*, p. 3):
>
> - sustaining a forward presence;
> - establishing and maintaining blockades;
> - deterring and defeating attacks on the United States, our allies, and friendly nations and, in particular, sustaining a sea-based nuclear deterrent force;
> - projecting national military power through modern expeditionary warfare, including attacking land targets from the sea, landing forces onshore and providing fire and logistical support for them, and engaging in sustained combat when necessary;
> - ensuring global freedom of the seas, airspace, and space; and
> - operating in joint combined settings in all these missions.
>
> These tasks are not new for the naval forces and have changed little over the decades. However, advanced technologies are now spreading around the world, and burgeoning military capabilities elsewhere will, in hostile hands, pose threats to U.S. naval force operations. The most serious are as follows (pp. 4-5):
>
> - access to and exploitation of space-based observations to track the surface fleet, making surprise more difficult to achieve and heightening the fleet's vulnerability;
> - increased ability to disrupt and exploit technically based intelligence and information systems;
> - effective antiaircraft weapons and systems;
> - all manner of mines, including "smart" minefields with networked sensors that can target individual ships for damage or destruction by mobile mines;
> - antiship cruise missiles with challenging physical and flight characteristics;
> - accurately guided ballistic missiles able to attack the fleet;

- quiet, modern, air-independent submarines with modern torpedoes; and
- nuclear, chemical, and biological weapons.

Future naval forces must be designed to meet these threats while maintaining the forward presence and operational flexibility that have characterized U.S. naval forces throughout history. This capability must be achieved in a world of ever-advancing technologies (particularly information technologies) available globally through the commercial sector and sales to foreign military users.

The study described the characteristics of future naval force operations as follows (p. 6):

- operations from forward deployment, with a few major secure bases of prepositioned equipment and supplies;
- great economy of force based on early, reliable intelligence; on the timely acquisition, processing, and dissemination of local and conflict- and environment-related information; and on all aspects of information warfare;
- combined arms operations from dispersed positions, using stealth, surprise, speed, and precision in identifying targets and attacking opponents, with fire and forces massed rapidly from great distances at decisive locations and times;
- defensive combat operations and systems, from ship self-defense through air defense, antisubmarine warfare, and antitactical ballistic missile defense, always networked in cooperative engagement modes that extend from the fleet to cover troops and installations ashore;
- Marine Corps operations in dispersed, highly mobile units from farther out at sea to deeper inland over a broader front, with more rapid conquest or neutralization of hostile populated areas, in the mode currently evolving into the doctrine for Operational Maneuver From the Sea;
- extensive use of commercial firms for maintenance and support functions; and
- extensive task sharing and mission integration in the joint and combined environment, with many key systems, especially in the information area, jointly operated.

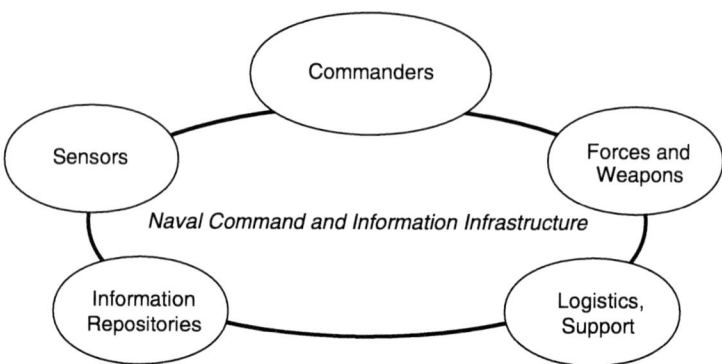

FIGURE 5-1 Diagram of the recommended structure for a naval, network-centric operational system (from National Research Council, 2000a).

deployed today that collect data that could be incorporated or "ingested" into a METOC data stream. Network technologies enable METOC to leverage other naval and joint battlegroup sensors and information. The command and control structure is hyperlinked, and hyperlinks are enabling interactive systems architectures, real-time adjustments to operating conditions, and rapid feedback on products and services.

Network-Centric Changes to the METOC Business Model

The end consequence of a network-centric business model is the movement away from stovepipes, away from hierarchical concepts of operation, tiered organizations, and closed processes toward an integrated network-centric working environment. This in turn requires that traditional information flow be modified so that the infrastructure supports the warfighter directly. These concepts are well expressed in the DOD (2001) definition of FORCnet:

> FORCnet is the architecture of warriors, weapons, sensors, networks, decision aids, and supporting systems integrated into a highly adaptive, human-centric, comprehensive maritime system that operates from seabed to space, from sea to land. By exploiting existing and emerging technologies, FORCnet enables disbursed human decisionmakers to leverage military capabilities to achieve dominance across the entire mission landscape with joint, allied, and coalition partners. FORCnet is the future implementation of network-centric warfare in the naval services.

With the evolution of network-centric warfare and the increased availability of environmental information from distributed sources, the existing METOC business model for meeting one critical mission objective—enhanced warfighting

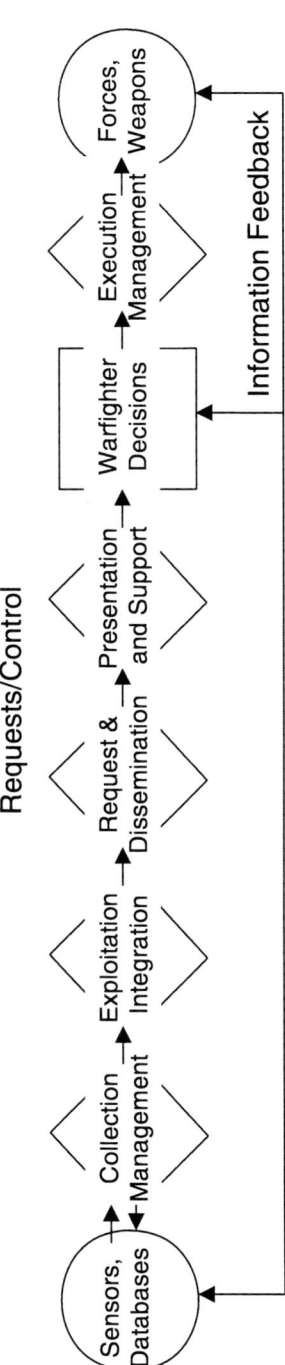

FIGURE 5-2 Functional architecture of the Naval Command and Information Infrastructure, as recommended by the National Research Council (National Research Council, 2000a).

capabilities—will become obsolete as the current business model does not adequately address the warfighter's needs. It does not facilitate close connectivity and fluid information exchange with the warfighter. It is difficult, therefore, to estimate the contribution of METOC warfighting products to improved capabilities.

The METOC business model for meeting the enhanced warfighting capabilities mission objective should be examined in light of e-commerce principles and network-centric warfare operational concepts. As discussed in Chapter 2, e-commerce principles reflect the changed relationship between a provider and his or her customer or end user, which has resulted from network principles introduced by Web technology. Customers (in this case warfighters) and service or information providers (in this case METOC) are now able and expect to have a dialogue about the nature of the product. Customization is now limited by the production process, not the customer interface. This review of the METOC business model should, therefore, be undertaken at three levels: customer interactions, networked sensors, and information fusion.

Two Japanese ships steam alongside Kitty Hawk during a Photo Exercise. Japanese ships integrated into Kitty Hawk's battle group during Exercise Keen Sword 2003. Keen Sword 2003 is the seventh in a series of regularly scheduled joint/bilateral field training exercises since 1986 involving the Japanese Maritime Self-Defense Force and U.S. military. The purpose of Keen Sword is to train and evaluate wartime functions and bilateral cooperation procedures against the backdrop of a regional contingency scenario that has direct and immediate consequences to the United States and Japan. Supporting multinational naval forces presents significant challenges and opportunities to the U.S. Navy METOC community (Photo courtesy of the U.S. Navy).

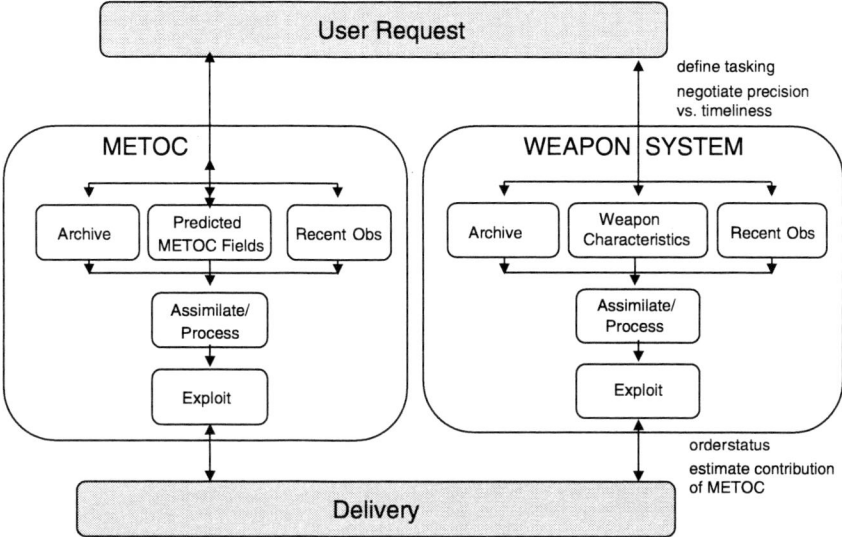

FIGURE 5-3 Level 1 adds a two-way communications loop for the user to begin interacting with the process to improve timeliness, accuracy, and relevancy of the METOC product. Information from the weapons system becomes available to the METOC provider.

LEVEL 1 RECOMMENDATION

Level 1 adds a two-way communications loop to the systems architecture of METOC warfighting support products (see Figure 5-3). Today, the warfighter sees only the tasking and the report or product. The purpose of two-way communications is to improve the warfighter's ability to influence and use existing products. By adding a two-way communications loop, the warfighter or asset commander can begin interacting with the process to improve timeliness, accuracy, and relevancy of the METOC product. Correspondingly, information from the warfighter/weapons system becomes available to the METOC provider. By adding chat rooms and software assistants between customers and the METOC provider, feedback is accommodated and user confidence improved. Feedback will also help synchronize data collections and assist the warfighter to assess his need for additional tasking. Finally, the end-of-mission synopsis of communications should be used to capture interactions and mine them.

LEVEL 2 RECOMMENDATION

The Level 2 system recommendation is to add methods for a distributed sensor network to the METOC business model (see Figure 5-4). This will increase

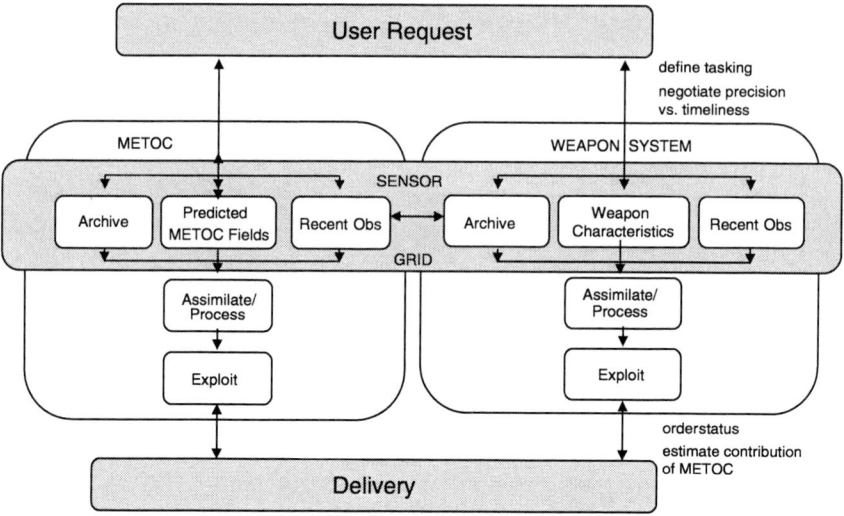

FIGURE 5-4 Level 2 adds a distributed sensor network and allows METOC access to data that were previously unavailable. It capitalizes on Internet Protocol/Transmission Control Protocol (IP/TCP) technology to address sensors across the battlefield.

opportunities for multipurpose use of data collected by other platforms. It will also improve data interoperability and open the door for METOC to access and ingest data that were previously unavailable. This has the potential to benefit METOC by adding temporal and spatial resolution to the ingested data stream. Its implementation capitalizes on Internet Protocol/Transmission Control Protocol technology to address sensors across the battlefield.

Network-centric warfare, as it applies to environmental support systems, is another way of achieving a "massively parallel" system. Such a system demands that considerable discipline be imposed in systems design and protocol. While keeping the architecture as open and flexible as possible, it will be necessary to define the functionality of individual nodes in the system and to provide identification on individual packets of information so that they may be used whenever and however they are received by other elements of the system.

LEVEL 3 RECOMMENDATION

The recommended systems implementation at Level 3 (see Figure 5-5) is a full breakdown of the stovepipes by adding sensor-, object-, and decision-level fusion to the exploitation process. This will greatly increase opportunities for finding answers that are unavailable with the sum of individual reports. It will

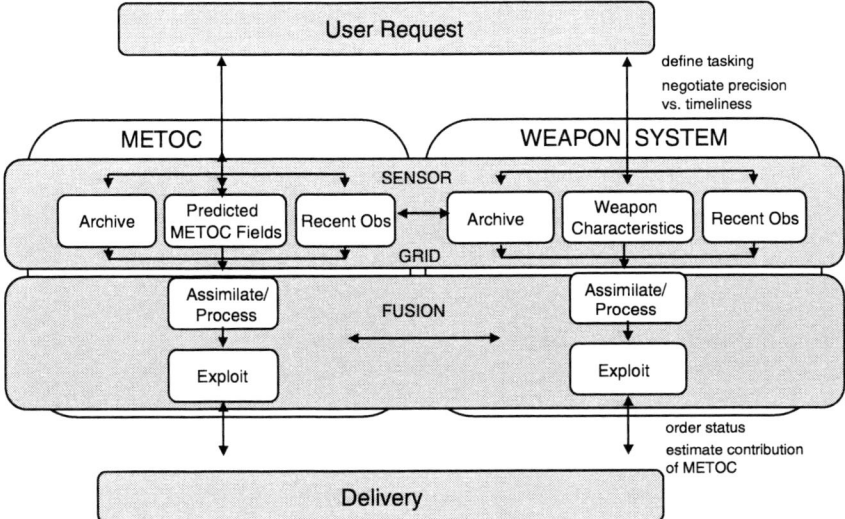

FIGURE 5-5 Level 3 adds a fusion and enables user to compute event outcomes in his or her own familiar framework. It also allows user to quickly draw inferences and suggest additional tasking and makes use of partially processed data as available.

enable a user to compute event outcomes in his or her own familiar framework. Breaking down the stovepipe, the user will be able to quickly draw inferences and suggest additional tasking. He or she will be able to make use of partially processed data as available and consolidate information. Fusion might be used to suggest additional tasking, inferences, and associations.

BENEFITS OF IMPROVED INFORMATION FLOW: AN EXAMPLE FROM UNDERSEA WARFARE

Chapter 2 discussed the types and limitations of METOC information as applied to undersea warfare (USW). Although USW is only one of many mission areas for which network-centric operations will enable better interaction between METOC and its warfighter, it is discussed here in greater detail to provide a concrete example. The following section, therefore, discusses how some of these limitations may be addressed, in part, by improved information flow via tactical decision aids (TDAs) intended to aid operational commanders. For passive systems there are two major environmental concerns for estimating SE (signal excess) for a sonar: TL (transmission loss) and beam-level ambient noise. Predictions of TL are now done with SFMPL and PCIMAT or with narrowband solutions to the wave equation that can handle range dependence. The environmental database is supported by MODAS and geoacoustic products such as GDEM. Ambient noise levels use several databases—HITS for shipping densities, ANDES, and DANES. While good for regional predictions in deep water, ambient noise can be very dynamic if near shipping and fishing, local biological noises, and weather-induced levels. Active systems rely on scattering strength measurements.

The fundamental question is how environmental information and TDAs are now used. Summarizing the experience of many users, including officers, it must be stated that there is not a high degree of confidence in the predictions because they often do not agree with the measurements. There are several possible reasons for this:

- The current TDAs are not robust and do not handle uncertainty and sensitivity to environmental data well. Robustness, sensitivity, and uncertainty are basic research issues and not simply a question of more detailed surveys.
- SFMPL output and PCIMAT output are not easy for a nonacoustician to use and understand.
- There are acoustic propagation modeling issues, especially for bottom-interacting paths and reverberation.
- There are gaps in the databases for important regions.
- Ambient noise models are not adequate.

Another concern is that uncertainty is not adequately captured or portrayed; thus, in many instances the operator will develop a false sense of security when using them.

Once predictions are proven to be unreliable, overall confidence is eroded. As a result, operators may lack confidence in accurate predictions, where environmental conditions and adequate data yield predictions with acceptable low degrees of uncertainty. Prediction codes are just one of many components in the decisionmaking process. Nevertheless, if progress is to be made in exploiting environmental data, this issue must be addressed as the operator views it (i.e., through his sonar) and not as a tabulation of environmental variability. Acoustic uncertainty is what really matters for sonar, not environmental variability. Presently, there is a very large gap between what the Naval Oceanographic Office (NAVOCEANO) perceives as the capabilities of its prediction tools and what the users and operators perceive. The interactive and parallel communication that lies at the center of network-centric operations would greatly enhance the role of user feedback, even on submarines, where interactions must, by their nature, be intermittent.

For submarine operations there are several issues where environmental information is not being provided or used as well as it could be. The first is that in discussing this with submarine commanders there seems to be a disconnect between what NAVOCEANO perceives and what the submarine commanders perceive. NAVOCEANO provides predictions that are stated to be good to ±2 dB; however, operators, based on practical experience, do not share this level of confidence. Some of this can probably be attributed to the cost metric used for establishing the fidelity of TL in the prediction codes.

Use of environmental information is two sided. First, TDAs, which can be used easily and which contain some insight into the complexities and uncertainties of acoustic propagation, are needed. Producing a number for TL is not very useful in itself. What is useful is to understand how it can be changed by how a submarine commander operates his submarine. On the other hand, submarine commanders need to invest time for feedback with NAVOCEANO and devote time for training in the use of the tools. Right now there is an unacceptable gap between the two.

One glaring omission in these bases is that data from other Navy sources (e.g., Office of Naval Research [ONR] experiments are not routinely used, even though they probably have the best environmental controls when done). There are many issues yet to be resolved before there can be high confidence that a beam output that the submarine uses is well predicted, even if TL is only approximately correct.

The importance of improving prediction tools cannot be understated. Recommendations include the following:

- The gap in perceived expectations must be reconciled. This can only be done with joint participation by NAVOCEANO and OPNAV, with the emphasis on actual sonar performance attained and that predicted.

- The problems of robustness, sensitivity, and uncertainty need attention from NAVOCEANO and within both the basic and applied research communities of ONR.

- There needs to be much greater awareness of the chain of oceanography to acoustics to signal processing in the development of TDAs. OPNAV is replacing the legacy sonars with modified commercial off-the-shelf acoustics (ARCI) systems.[1] Future environmental information needs to match the requirements of the ARCI systems.

- Methods for incorporating the outputs of in situ sonar to provide feedback on the quality of TDA predictions need to be developed.

---

[1]Acoustic rapid COTS (commercial-off-the-shelf) insertion, or ARCI, is a modernization effort designed to bring about rapid improvement in processing performance at low cost by modifying and then installing existing commercial off-the-shelf technology to improve acoustic sensor performance onboard submarines. While using the same sonar arrays, ARCI has demonstrated significant improvements in the ability of ARCI-equipped submarines to detect other submarines. ARCI is the baseline sonar system for the Virginia Class SSN and is designed to be retrofitted to existing submarines as part of a four-phase program.

• Address the problem of the appropriate scales of environmental data needed for operational sonars.

• Develop methods for archiving original acoustic and environmental data acquired from ONR experiments. (This is fundamental especially in areas such as the South China Sea where NAVOCEANO's access will be limited.)

SUMMARY

An analysis of the current information flow between the METOC community and its customers suggests there are areas where the existing information flow appears to be inadequate or mismatched to user needs. Based on this and previous National Research Council studies, it appears that, by leveraging network-centric business principles in these areas, it is likely that uncertainty can be substantially reduced and/or that accuracy and timeliness can be improved. The greatest benefits of network-centric business principles are derived when (1) improved spatial and/or temporal resolution can substantially reduce uncertainty, (2) confidence in the outcome can be improved through information fusion, (3) the METOC process can rapidly respond to additional or modified tasking, and (4) the risks and uncertainties in outcomes can be stated in the user's framework.

Two major findings can be derived from the information flow analysis. First, the current METOC information flow for its two primary missions, Safe Operating Forces and Enhanced Warfighting Capabilities, is generally a one-way stovepipe; the customer requests and receives a product but has no intermediate data access or visibility into the process.

This one-way stovepipe process works reasonably well for delivery of global gridded model output and for general forecasting services derived from model output. However, for a variety of reasons, stovepiped information flow is substantially less effective for products designed to enhance warfighting capabilities. Warfighters and weapons systems require tailored products, and they require them rapidly and/or at higher spatial resolution, especially when changing military and environmental conditions need to be predicted on a compatible timescale.

The second finding is that there appears to be an impedance mismatch at the interface between METOC products designed to enhance warfighting capability and the end-user weapons systems. Although responsibility for this mismatch may rest with the warfighter, it is recommended that the METOC community take aggressive action to address it. If left with the warfighter, a solution may be developed that does not optimize the use of environmental information to reduce the warfighter's uncertainty.

It is recommended that network-centric business principles be applied to the METOC information flow for products intended to enhance warfighting capabilities. Network-centric operations will enable uncertainty of outcomes to be quantified and to be presented to the end user in his context. Network-centric operations will also enable uncertainty to be reduced by increasing data sources, enabling fused decision logic and integrated rapid response to changing environmental and military conditions.

As discussed in *Network-Centric Naval Forces*, the implementation of network-centric operations does not start from a zero base. The naval forces are faced with transforming today's systems—including "legacy" subsystems, new ones entering service or under development for future service, and also elements of subsystems of other services, national agencies, and possibly coalition partners—into new, all-inclusive systems. All of these subsystems and their components must be able to operate together, even if they were not originally designed to do so. All must be accounted for in devising network-centric concepts of operation and in designing the systems that will support them.

One of the greatest problems in shifting from today's platform-centric operational concepts to tomorrow's network-centric operational concepts is being able to ensure interoperability among the subsystems and components of the fleet and the Marine forces as well as joint and coalition forces. The forces can operate to their full potential if all subsystems and information network components can operate smoothly and seamlessly together. In the current context "interoperability" does not necessarily mean that the characteristics of all subsystems and components must match at the level of waveforms and data formats. Interoperability means that the subsystems must be able to transfer raw or processed data among themselves by any means that can be made available, from actually having the common waveforms and data formats to using standard interfaces or intermediate black boxes enabling translation from one to another.

Ensuring interoperability will be a very complex and technically intensive task involving network protocols, data standards, consistency algorithms, and many other aspects of networking design as well as numerous procedural matters. The subsystem mix will evolve and will be different from the one that exists today.

6

Moving Ahead

> This chapter integrates points from the preceding chapters and points out that:
>
> - prioritizing data acquisition should follow a structured decision process that emphasizes least-cost methods for reducing the impact of environmental uncertainty;
> - leveraging existing efforts outside the naval meteorological and oceanographic (METOC) enterprise to collect environmental data offers many advantages to committing additional resources to expand primary data collection efforts;
> - the ongoing Department of Defense (DOD)-wide transition, including expanded focus on network-centric operations, will provide both challenges and opportunities to the naval METOC enterprise; and
> - embracing principles associated with network-centric warfare and decisionmaking in the face of uncertainty would position the naval METOC community to better support U.S. Naval Forces in the 21st century.

June 2004, Indonesia: A ship suspected of carrying contraband arms has been tracked and is heading for the harbor of a hostile nation. A SEAL team is to be dispatched by high-speed boat to make the interception and to board and take control of the vessel. The team leader checks with the METOC officer and ascertains that the wave heights are averaging 3 feet and the wind is from the south-

west. He estimates the intercept can be made a few miles before the ship enters hostile waters if the boat average 45 knots. Having second thoughts, the METOC officer consults his records and confers with his staff. A senior petty officer is able to search online through a broadband communications link and finds cruise reports from several battle groups that had previously operated in the area. Two of the reports noted that in June abrupt shifts in wind speed and direction during evening hours are not common. He then consults a forecast for the region posted to the Web by the European Centre for Medium-Range Weather Forecasts that suggests a wind shift may be likely. An online consultation with the Fleet Numerical Meteorology and Oceanography Command (FNMOC) and the Naval Oceanographic Office (NAVOCEANO) ensues, and a new local forecast is created based on a recently released, nested model that incorporates far-field wave and swell data. The METOC officer contacts the SEAL team leader and explains that an unfavorable change in sea state is likely and could make it impossible to make speeds over 40 knots. They agree that the mission should commence two hours earlier just to be safe. The intercept is made successfully and on schedule.

May 2005, Gulf of Oman: A nuclear carrier (CVN) and three cruisers choose to anchor for the evening in the Gulf of Oman in the lee of Masirah Island. The METOC officer tasks his division with searching its library of cruise reports with searching online for any information on the Masirah anchorage. He directs that an e-mail be sent via a broadband connection to NAVOCEANO, located at the Stennis Space Center in Mississippi. The searches are fruitful. Previously compiled cruise reports have been retained online and are carefully reviewed. The Naval Oceanographic Office also transmits the latest Special Tactical Oceanographic Information Chart that covers the anchorage. All of these sources point out that the anchorage is near a major ocean upwelling and is teaming with marine life. There is a particular hazard from jellyfish that are attracted to the surface at night by bright lights. It is noted that these jellyfish can clog vital machinery and cause critical damage. The proper action is to ensure that "darken ship" is implemented on all ships. This means all hatches, doors, ports, and other openings must be made light tight and that no bright lights are allowed topside. The METOC officer reports his findings to the carrier executive officer, and the proper orders are given. The anchorage is completed successfully.

February 2006, Somalia: An important naval gunfire support mission is called for during the early morning hours to support U.S. Marines deployed inland as part of a multinational force committed to breaking up a concentration of hostile combatants with ties to an international terrorist organization. The weather forecast predicts clear skies. The embarked METOC officer is not convinced the weather over the coastal targets will be clear enough for accurate gunfire visual spotting. He consults with his division and establishes broadband communications with the Fleet Numerical Weather Facility in Monterey and the

Naval Oceanographic Office in Mississippi. An online chat session is established, and the climatological records are accessed. Forecasters, including a METOC officer with experience in Somalia, agree that an early morning temperature inversion is not unusual for that time of year. Coupled with the smoke from wood-burning fires and stoves common in developing nations, this situation can obscure visibility if winds are not present. Based on this information, the METOC officer directs that an unmanned airborne vehicle (UAV) be launched over the target area at sunrise. The optical UAV and METOC sensors, including a dropsonde, verify that visibility will be poor and visual spotting of rounds on target will be very difficult. The possibility of collateral damage will be great. The operational commander consults with senior commanders in the joint operating force and delays the operation and the gunfire support mission until 1000 hours, when the inversion layer has lifted. The mission is successfully conducted at that time.

July 2006, Washington, D.C.: *The technical director of the Office of the Oceanographer of the Navy reviews a report that makes a convincing argument for expanding efforts to systematically identify non-DOD websites with significant environmental information. The Office of the Oceanographer works with the Office of Naval Research (ONR) to initiate a program to develop "smart" Internet search programs that can automatically and rapidly search millions of websites, identify resident databases, import and reformat the data, and make it available for METOC use. The primary argument used in the report was based on a statistically rigorous analysis of three years' worth of Internet relay chat (IRC chat) messages in which a significant number of problems posed by forward-deployed METOC personnel were solved when personnel at FNMOC or NAVOCEANO obtained additional data or information from a non-DOD site.*

Understanding the distribution of friendly, enemy, and neutral forces and facilities and the nature and significance of the environment they occupy is a key component of what has been described by the U.S. Navy as battlespace awareness. A shared awareness of the battlespace among allied military forces is considered to be a major advantage and a force multiplier that is recognized as both highly desirable and difficult to achieve. Before data can be shared, data must first be obtained and then rendered into a usable—form, in other words, information or knowledge. Efforts to collect, assimilate, analyze, and disseminate information about and predict the nature, distribution, and intent of enemy forces have long been the focus of a large and complex intelligence, surveillance, and reconnaissance effort. Efforts to collect, assimilate, analyze, and disseminate information about and predict the nature and significance of the environmental character of the naval battlespace, though less well known, have been the focus of a complex meteorological and oceanographic effort referred to within the U.S. Naval Forces as METOC.

U.S. Marines assigned to the 24th Marine Expeditionary Unit (MEU) Special Operations Capable (SOC) disembark from a Landing Craft Utility (LCU) 1662 assigned to Assault Craft Unit Two (ACU-2). Marines from the 24th MEU are conducting exercises in the U.S. Central Command Area of Responsibility while on a regularly scheduled deployment in support of Operation Enduring Freedom. Support for expeditionary warfare requires close cooperation among the various military services and DOD agencies (Photo courtesy of the U.S. Navy).

The naval METOC enterprise is a complex system of platforms, personnel, and computer systems designed to support operations carried out by U.S. Naval Forces by producing high-quality tailored environmental information products (see Chapters 2 and 3 for greater detail). As pointed out by the scenarios above, the intended consumers of this information include decisionmakers facing a variety of complex choices, some of which may be significantly affected by environmental processes operating at a variety of temporal and spatial scales. As also pointed out in the previous chapters and in these updated scenarios, new approaches to acquiring, managing, and disseminating environmental information to U.S. Naval Forces offer many opportunities for improved tactical decisionmaking.

The current system, however, is limited by what appears to be a lack of performance metrics, including a robust understanding of how platforms and personnel are affected by environmental processes (see Chapter 3). Valid and

> **BOX 6-1**
> **METOC and the U.S. Marine Corps**
>
> A brief survey of the Marine Corps METOC capability has shown that a Navy-like evolution from pure aviation weather support to a more extensive program including surface weather and oceanographic forecasting is taking place. Unfortunately, the Marine Corps is many years behind the Navy and Air Force largely due to a lack of personnel, material resources, and educational opportunities.
>
> The U.S. Marine Corps is about one-half the size of the U.S. Navy but has one-tenth the officer METOC resources. The U.S. Navy METOC organization has over 1,700 members, whereas the Corps has about 450. Virtually all of the 40 Marine Corps METOC officers have worked their way up through the ranks from weather forecaster to larger leadership roles. The most senior METOC billet is that of a lieutenant colonel (O-5) and is presently being filled by a major (O-4). METOC departments are normally subordinated to intelligence departments. The U.S. Navy presently has two METOC specialty admirals on active duty, a vice admiral, and a lower-half rear admiral. Additionally, the present Oceanographer of the Navy, who comes from a warfighting community, is an upper-half rear admiral who actively champions METOC issues as a member of the staff of the Chief of Naval Operations.
>
> The U.S. Marine Corps has no formal education program for its METOC officers. Most officers do not have a bachelor's degree unless they earned one in an off-duty program on their own. Almost all of the

quantifiable feedback from warfighters and other operators, while difficult to obtain, is needed if objective criteria for data acquisition are to be established. Limited funding, limited time, and the rapidly evolving nature of naval and expeditionary warfare make sound decisions regarding data acquisition and dissemination a high priority of the naval METOC community.

As discussed in Chapter 3, the spatial and temporal scales of various environmental processes may have a profound effect on efforts to predict future environmental states. Areas where inadequate predictive skill is of concern require greater numbers and more recent observations, either to support data analysis for prediction or to supplant predictions altogether. Determining what level of accuracy is needed in predicting future states of some environmental phenomena is limited by a clear understanding of the sensitivity of platforms and personnel to various environmental conditions. Thus, before rigorous efforts at setting data acquisition priorities can be completed, a fuller account of the impact of various

> 400 officers in the Navy have a college and usually two master's degrees in oceanography and/or meteorology—a requirement for anyone aspiring to success within the community. The Marine Corps officers the committee interviewed are top-notch performers, and they seemed eager to gain additional education if it were offered.
>
> The U.S. Marine Corps follows U.S. Navy standards for equipment at its installations but has very few lightweight deployable systems to take onshore early in an assault. Heavy METMP(R) vans have been developed to support established air operations onshore, but they cannot be easily lifted to shore in the first stages of an assault. For example, no vans were lifted into Kandahar during the war in Afghanistan. The vans also require significant training and support since they can be equipped with up to five different computer operating systems.
>
> The U.S. Marine Corps has relied on U.S. Navy METOC support, but the situation is changing. During operations in Afghanistan the radio links back to the ships traversed 500 miles, and communications were limited at best. There is a clear desire to have more METOC support internally and to rely less on the U.S. Navy. Forward-deployed weather and coastal modeling systems have great appeal, and the Corps is becoming more active in these areas. These efforts are constrained by limited support for equipment development and deployment as well as low levels of METOC-expendable supplies. In the current international political environment it is clear that new approaches are needed to provide assault units with deployable and dependable organic METOC support.

processes on naval operations must be acquired. The General Requirements Database is a good first step, but greater and more clearly defendable establishment of critical thresholds, including compound effects from multiple environmental processes, is needed. With this information in hand, efforts to weigh the benefit of additional information against the cost of acquiring it could be undertaken. Prioritizing data aquisition should follow a rather simple adaptive management process similar to this:

1. Establish mission-critical environmental thresholds for all platforms (e.g., a comprehensive list of environmental processes or phenomena of concern along with numerical values for the threshold that each would impair or severely degrade mission success).

2. Identify and evaluate all relevant environmental data presently collected and readily accessible to the METOC community.

3. Develop a numerical measure of uncertainty that can be applied to all processes or phenomena of concern identified.
4. Develop a METOC decision aid that would prioritize environmental information needed for any given naval activity, based on the mission and platform involved.
5. Determine the least-cost alternative (e.g., change platform, harden platform, acquire additional data) for reducing uncertainty to an acceptable level.

As is often the case, developing the characteristics of a process is much more difficult than developing the information needed to successfully complete each component task. Significant effort should be redirected toward establishing this prerequisite information. Such redirection of effort will largely involve a re-examination of how the needs of U.S. Naval Forces environmental information are established.

Developing More Effective METOC Processes in the Near Term: Leveraging Existing Programs and Resources

Information about environmental conditions is developed from observation or inference. Understanding the nature of environmental processes provides the skill needed to determine when additional observations are needed or to build on existing information to draw inferences about conditions in the future. This latter process is at the heart of traditional forecasting and has been greatly expanded by advances in the capability and capacity of computing facilities and remote sensing. Much work, however, remains to be accomplished if forecasting is to achieve the accuracy and reliability needed at the temporal and spatial scales relevant to many naval operations, especially those taking place in coastal areas.

It will not be possible to obtain, manage, and disseminate environmental information at all scales of interest for all areas of possible naval activities in the foreseeable future. Like many entities with an operational focus, the naval METOC enterprise has evolved to operate on very short production cycles. At present, different information for various geographic areas of concern and different portions of the battlespace (see Figure 6-1) is distributed over multiple sources, many of which are identified by various METOC officers "in stride" as reports are developed for various customers in response to real-time requests. Success in this approach is largely dependent on the knowledge, creativity, and experience of individual METOC officers. Lack of a more cohesive proactive approach to priority setting limits the METOC community's ability to identify, evaluate, and acquire data and information from nontraditional sources during emerging crises. The METOC community needs to become both a supporter of network-centric operations and a beneficiary of those operations by being an active user of the networks being developed to support them. **The METOC enterprise should incorporate network-centric approaches to enable easy and flexible**

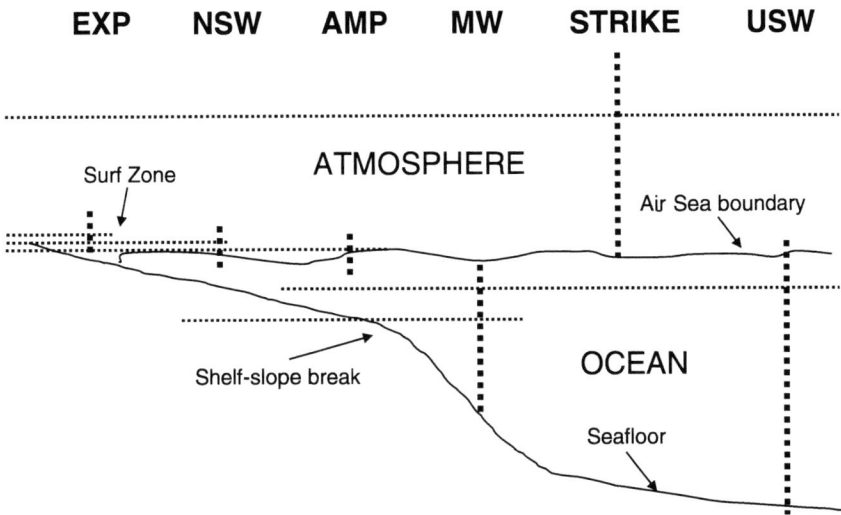

FIGURE 6-1 Stylized cartoon showing distribution of some warfare mission areas relative to major environmental boundaries or gradients. Note that many warfare mission areas occur exclusively above, below, or along the air-sea interface.

interconnectivity among the individual METOC officers and with nontraditional sources of information. This should be done now to leverage the people and knowledge assets currently in place.

As discussed in Chapter 4, special attention should be given to identifying the METOC contribution for nonroutine operations (e.g., evacuation of noncombatants, amphibious warfare, as opposed to activities such as ship tracking or air operations that occur on a daily basis and thus tend to be continuously evaluated and modified). Guidance for identifying broadly needed and significant information across multiple warfare areas should be derived from an understanding of the benefit of additional information for reducing uncertainty versus the cost of improving the content and reliability of environmental information, whether through additional observations, improved understanding of the underlying physical processes, or more powerful forecasting tools that take advantage of both (see Chapter 4 for fuller discussion).

The availability of unmanned airborne vehicles (UAVs) and unmanned underwater vehicles, and their expanded capability to covertly collect intelligence, surveillance, and reconnaissance information in denied areas using a variety of electrooptical and acoustic sensors, creates a largely untapped potential for the unintended use of such information to support the development or validation of METOC products or forecasts. **The Oceanographer of the Navy and the Com-**

mander of Naval Meteorology and Oceanography Command (CNMOC) should work with the broader community within DOD and elsewhere to expand efforts to make intelligence, surveillance, and reconnaissance information and data with environmental content more accessible to the METOC community while protecting sensitive sources.

Such efforts should include expanded efforts to remove unneeded or particularly sensitive nonenvironmental content; thus, reducing security risk while making the environmentally relevant information or data acquired during intelligence-gathering, surveillance, and reconnaissance efforts more accessible to the METOC community. At the same time, the METOC community's ability to securely handle sensitive georeferenced material must be expanded. **The METOC community should also seek to acquire its own unmanned platforms since concerns about security can limit the availability of such collateral information. In addition, it is strongly recommended that ONR evaluate the potential for exploiting existing intelligence-gathering, surveillance, and reconnaissance sensors as dual-use METOC sensors. Assuming this potential is significant, the Oceanographer of the Navy and CNMOC should work with the DOD community to develop mechanisms to exploit this potential.**

Particular emphasis should be placed on forward-deployed assets that are already under the control of theater commands. Data collection could be either specifically tasked or while en route to other missions. Analysis could be easily handled onboard. Care must be taken that the primary missions of the intelligence, reconnaissance, or surveillance operations involved are not hindered or compromised so that such dual-use activities are indeed cost effective.

As discussed in Chapter 5, as network-centric warfare becomes an operational reality, enhanced computing and communications capabilities are changing the way U.S. Naval Forces fight, communicate, and plan. Extensive e-mail and METOC electronic chat room traffic is already overtaking the formal Naval Message System and creating peer-to-peer linkages that are a radical departure from the hierarchical system that has been in place for years. Faster computers and high-speed data links resulting from the IT-21 (Information Technology 21st Century) initiative and the Navy-Marine Corps Internet program are accelerating this dramatic change in the METOC community.

The Office of the Oceanographer of the Navy and CNMOC should work with regional METOC commands to formally define a network-centric concept of operations that embraces peer-to-peer networking within the METOC community while preserving the flexibility and timeliness that have led to the rapid growth in its use. The goal of formalizing this type of exchange should be to improve information content and its usefulness as a source of insight into user and customer needs (e.g., opportunities for data mining, frequency of various types of information requests, identification of systematic problems in information access) while encouraging continued and wider usage. Access to, and the transmission of, METOC data between ships and to shore facilities will be significant parts of the network-centric warfare transformation. Current efforts to incorporate network-centric principles into METOC operational concepts are only beginning to tap the vast potential of the network-centric operational concepts.

Expanding METOC Capabilities: Logical Next Steps

Existing capabilities for data collection, storage, and dissemination can produce voluminous bodies of information with varying amounts of useful content. The sheer volume of information is already posing an unforeseen challenge as the METOC community and the warfighters they support struggle to match useful information to key users. **The Oceanographer of the Navy should ensure that CNMOC, FNMOC, and NAVOCEANO jointly develop a strategic plan for data acquisition over the next 10 to 20 years that prioritizes geographic regions of focus, incorporates an understanding of the limits of environmental information currently available to the METOC community, and evaluates such technologies as distributed databases, advanced information data-mining techniques, and intelligent agent technology.**

In addition to geopolitical considerations (which may be fairly fluid on decadal timescales), such a plan should be based on a thorough understanding of what environmental information is currently held or available and which processes or conditions will be of particular concern to various naval missions. **Once an initial framework is established, CNMOC and NAVOCEANO should work with operational commanders to evaluate the adequacy of existing critical information (e.g., external variables such as bathymetry/topography, sediment type/land cover) and plan for filling data and information gaps.**

Increasing bandwidth, while relaxing some constraints, will undoubtedly lead to further dilution of information content. As the locations of Navy and Marine actions vary across multiple continents and adjoining seas, greater effort must go into developing mechanisms for rapid and efficient environmental characterization that focuses on providing the warfighter with targeted information with a high proportion of useful content while minimizing ancillary or irrelevant information. High-quality data are a must if the quality and utility of information are to be high. **The Office of the Oceanographer of the Navy should foster efforts (by providing expertise and access) by ONR to develop and implement a system that promotes optimized environmental characterization, keyed to action-specific warfighter needs during various naval missions or suites of missions.**

A significant component of the METOC enterprise, in terms of both fiscal and human resources, is devoted to data collection. Understanding how new data collection, as opposed to use of archived data or numerical extrapolations or interpolations, improves the content of environmental information (i.e., reduces uncertainty) should be a key component of targeted data acquisition. Since data collection resources are limited, and because the cost of data acquisition in denied areas can be very high, methods for establishing data collection efforts and the research and development that support data collection platform development, should be focused using objective criteria. **The Office of the Oceanographer of the Navy and CNMOC should invest in the development of formal and rigorous methods for identifying high-priority data needs that are specific to the platforms and missions to be involved in any potential naval action. Determination of the cost of uncertainty, and focusing on data collection efforts that result in the greatest reduction in total cost of uncertainty, should be given priority at ONR and the Office of the Oceanographer.**

Asset allocation should be based on achieving improvements in the most significant parameters of interest. Such an effort will need to be based on a rigorous understanding of critical thresholds for platforms and systems involved as well as the spatial and temporal variability of key parameters and the operational tempo associated with each mission or suite of missions.

At present, there is insufficient continuity of responsibility and feedback for maintaining databases and models. In addition, there is inadequate exchange of data or information collected or managed throughout the Department of the Navy. **The Oceanographer of the Navy should clarify the various areas of responsibility and assess the performance of such databases, models, and the tactical decision aids, focusing on their value for individual mission areas.** In other words, the value or adequacy of a specific database or model may vary by mission or project, but information collected for a specific mission or project may still be of value to unintended users. Furthermore, at present there is insufficient use of datasets collected by other federal agencies and academia. **CNMOC and FNMOC should expand efforts to identify data of value and work with ONR to develop methodologies for evaluating and bringing data into existing METOC systems.** Once expanded capabilities to access data and information from a variety of sources, whether from within the DOD, academia, or other nongovernment sources, is established, an effort should be made to develop and implement a system that permits rapid retrieval of environmental data collected in specific geographical areas. Since many METOC products are intended to aid in optimizing the performance of individual weapons systems, provisions should be made to incorporate METOC into weapons systems so that the products are sufficiently integrated to effectively inform and guide operators of the systems (and their supervisors).

The traditional role of the Marine Corps in expeditionary warfare and the renewed focus on littoral operations involving the U.S. Navy continue to drive the need for environmental information in coastal areas where access is frequently denied. Efforts to improve secured, low-profile communications, to alleviate the risk to troops and Marines in coastal areas from chemical and biological agents (either from the tactical deployment of weapons of mass destruction by enemy forces or the destruction of such weapons by friendly forces), and to provide accurate assessments of atmospheric conditions during the planning and implementation of strike missions have placed a greater emphasis on coastal meteorology. This includes development of predictive meteorological capability at fine scales and intensive data gathering in coastal environments. Military operations in these areas require data in forward operations areas. UAVs are ideally suited for this and provide additional opportunity for data gathering on behalf of METOC. **ONR should develop, and the Oceanographer of the Navy should work with operational commands to deploy, atmospheric sensors on UAVs that will permit the collection of essential environmental information without impairing the intelligence, reconnaissance, or surveillance efforts they are largely designed to carry out.**

Current efforts to model many littoral processes of importance are promising but are not fully operational. Thus, needed predictive capabilities are not currently available. **The Office of the Oceanographer should foster research and development efforts at ONR to integrate mesoscale models with smaller-scale local littoral models.**

Changing Attitudes and Approaches: A Longer-Term Vision

The philosophy and approaches used by the Navy's oceanographic community to supply METOC information to the fleet and Marine Corps can be described in terms of a business model (in fact, the Navy and Marine Corps, like the DOD and most federal agencies, have already adopted this philosophy to some degree, as testified by widespread use of such terms as "user driven" and "customer needs" in planning documents). The present METOC business[1] model for providing global and mesoscale forecast fields is well defined and successful. However, the METOC business model for providing enhanced warfighting capabilities does not adequately address customer needs. The current relationship does not facilitate close connectivity or fluid information exchange with the customer. Therefore, it is difficult to quantify the value of environmental information or its impact on warfighting.

It is unclear whether the existing METOC enterprise reflects a strategic or unifying principle that can help the various component parts understand their relationship to one another and the overall goal of the activity. Providing specific advice for improving the overall performance of the METOC enterprise is therefore difficult. A review of various DOD and Department of the Navy guidance documents, however, does suggest that, in order to keep pace with changes now being undertaken by the operational Navy and Marine Corps, the METOC com-

---

[1]Use of the term "business model" may initially seem inappropriate to some readers in a report about supporting warfighters. This report, like other reports that discuss business models, uses the term "business model" to describe the mechanisms and underlying philosophies that characterize a serious and organized endeavor. Thus, its use here is not intended to equate conducting war with conducting commerce.

munity needs to reexamine many of its core approaches in a more systematic manner. **The de facto business model currently employed by the METOC community for providing enhanced warfighting capabilities should be examined and modified in light of e-commerce principles (e.g., peer-to-peer tasking and Web-enabled customer service) and network-centric warfare concepts of operations.** This review should be undertaken at four levels: customer interactions; data collection, data fusion, and information management; sensor networks; and network-centric operations.

To move forward and remain state of the art in environmental forecasting and prediction, resources need to be put into the development of both observational capability and models of processes on small time- and space scales. There is a critical need to improve feedback in data and prediction flow between the METOC community and its customers Failure to do so will compromise the ability of the METOC community to move forward and provide reliable predictions. **The METOC community must build on existing relationships to strengthen its ties to operational U.S. Naval Forces and to the academic community, especially in the area of data assimilative forecast models.**

The role of the METOC enterprise is to provide information about environmental conditions that may be encountered by and thus impact naval operations (including actions of expeditionary forces). In other words, the goal of METOC is to reduce uncertainty, gauged by its cost, to naval operations. This can be done on two levels. At the operational level that forms the primary focus for decisions made by the Oceanographer of the Navy as a resource sponsor or by CNMOC as the head of the METOC claimancy, naval needs are near term and problems are often immediate. ONR, on the other hand, is charged with developing a science and technology investment strategy for the longer term. The approaches appropriate for each endeavor are therefore quite different.

At an operational level, it may be possible to collect additional data that can better constrain boundary or initial conditions for numerical models, thus reducing uncertainty in predictions. The problem then becomes determination of the optimal collection strategy, including which instruments to deploy and where to sample. What is the use of resources that provide the greatest reduction of uncertainty (specifically the cost of uncertainty, which depends on the operational significance)? The appropriate answer depends on sensitivities of the model in question and the available sampling approach, and can be formalized through one of several business decision theories such as Bayesian statistics.

In the presence of perfect battlespace awareness, perfect tactical decisions are theoretically possible.[2] Perfect environmental information, however, is

---

[2]Obviously there are several barriers to making perfect tactical decisions. In situations where environmental conditions play a role, imperfect understanding or inaccurate information can contribute to poor tactical decisions. Perfect knowledge of the environment would then allow, but not guarantee, perfect tactical decisions.

neither achievable or even necessary in many instances. Thus, while uncertainty in environmental predictions introduces costs in terms of the increased risk and occurrence of failure or the costs of contingency plans or suboptimal tactics, the benefit of better information must always be weighed against the cost of developing it. The goal of the METOC community is the reduction of environmental uncertainty in mission planning and operations. The optimum investment strategy is that which reduces environmental uncertainty to the level necessary for a desired probability of mission success and no more. Measures of the cost of uncertainty are not linear but are strongly concentrated on critical thresholds and are weighted differently for different variables and also by the value of the mission. **The Office of the Oceanographer of the Navy should promote reduction in the cost of uncertainty as a measure of value, so that a sensible strategy for research and development investments can be developed.**

For the research programs at ONR, the investment is not in augmented data collection but in the development of fundamentally better approaches. In contrast to the more applied METOC problem, there is no simple transfer function to equate the cost of research with expected return in reduced uncertainty. Thus, business methods for determination of the optimum rate and direction of investment cannot be employed. Instead, the cost of uncertainty provides an objective measure of research needs among many processes of disciplines that can drive the directions of research. Management wisdom must drive the decisions of funding rates and potential payoffs. **ONR should continue to expand its efforts to understand and quantify both uncertainty and its cost in military operations. Furthermore, research priorities should incorporate an understanding of the relative impact uncertainty has on various naval operations, so that research priorities map to areas where the cost of uncertainty is the greatest.**

To more fully capture the benefits of improved measures of environmental uncertainty and the cost of that uncertainty, operational commanders need to more fully understand the accuracy of environmental information provided them. The current naval concept of operations for the understanding and assessment of environmental uncertainty is contained primarily in the collective experience of the METOC community, enhanced by informal and nondoctrinal infrastructures; thus, the quality and utility of environmental information will remain uneven and ephemeral, paced by the posting cycle of personnel. **The Office of the Oceanographer should work with the U.S. Naval Forces operational commanders to introduce and explain the concept of environmental uncertainty and its value (including development of a common nomenclature for expressing uncertainty). U.S. Naval Forces in general, and the METOC community in particular, should take advantage of the concept of environmental uncertainty in more formal and recognized ways.**

Cited References

Alberts, D.S., J.J. Gartska, and F.P. Stein. 1999. Network Centric Warfare—Developing and Leveraging Information Superiority, 2nd Ed. Department of Defense, CCRP Publication Series, Washington, D.C.

Avery, J., 1998. The naval mine threat to U.S. surface forces. Surface Warfare (May/June):4-9.

Berger, J.O. 1985. Statistical Decision Theory and Bayesian Analysis. 2nd Edition. Springer-Verlag, New York.

Chassignet, E.P., H. Arango, D. Dietrich, T. Ezer, M. Ghil, D.B. Haidvogel, C.C. Ma, A. Mehra, A.M. Paiva, and Z. Sirkes. 2000. Data assimilation and model evaluation experiments North Atlantic basin: The base experiments. Dynamics of Atmospheres and Oceans 32:155-183.

Chickadel, C.C., R.A. Holman, and M.F. Freilich. An optical technique for the measurement of longshore currents. Journal of Geophysical Research, in review.

*Department of Defense. 1994. Joint Publication 1-02, Dictionary of Military and Associated Terms.

*Department of the Navy. 1996. COMNAVMETOCCOM Concept of Operations.

*Department of the Navy. 1997a. Forward...From the Sea. Chief of Operations.

*Department of the Navy. 1997b. Naval Meteorology and Oceanography Command, Strategic Plan.

*Department of the Navy. 1999. Office of the Oceanographer of the Navy Strategic Plan C.

*Department of the Navy. 2000. Navy Strategic Planning Guidance/Strategic Planning Objectives.

*Department of Defense. 2000. Joint Vision 2020. Joint Chiefs of Staff.

*Department of Defense. 2001. Defense Planning Guidance, 2001.

*Department of the Navy. 2002. Naval Oceanography Program: Operational Concept.

Ferson, S., and R. Kuhn. 1994. RiskCalc^{TM}: Uncertainty Analysis with Interval and Fuzzy Arithmetic. Applied Biomathematics, Setauket, New York.

Ferson, S., and L.R. Ginzburg. 1996. Different methods are needed to propagate ignorances and variability. Reliability Engineering and System Safety 54:133-144.

Gelman, A., J.B. Carlin, H.S. Stern, and D.G. Rubin. 1995. Bayesian Data Analysis. Chapman and Hall, London.

---

*These publications can be found through www.dtic.mil.

CITED REFERENCES

Haidvogel, D.B., and A. Beckmann. 1998. Numerical models of the coastal ocean. The sea, the global coastal ocean, processes and methods, K.H. Brink and A.R. Robinson, (eds.), John Wiley & Sons, Inc., New York 10:475-482.

Haidvogel, D.B., H.G. Arango, K. Hedstrom, A. Beckmann, P. Malanotte-Rizzoli, and A.F. Shchepetkin. 2000. Model evaluation experiments in the North Atlantic basin: Simulations in nonlinear terrain-following coordinates, Dynamics of Atmospheres and Oceans 32:239-281.

Handfield, and T. Clark. 1999. General Requirements Relational Database.

Hobert, J.P., and G. Casella. 1996. The effect of improper priors on Gibbs sampling in hierarchical linear mixed models. Journal of the American Statistical Association 91:1461-1473.

Holland, K.T., B. Raubenheimer, R.T. Guza, and R.A. Holman. 1995. Runup kinematics on a natural beach. Journal of Geophysical Research 100:4985-4993.

Holman, R.A., and R.T. Guza. 1984. Measuring run-up on a natural beach. Coastal Engineering 8: 129-140.

Holman, R.A., T.C. Lippman, P.V. O'Nell, and K. Hathaway. 1991. Video estimation of subaerial beach profiles. Marine Geology 97:225-231.

Holman, R.A, T. Holland, H. Stockdon, and C. Church. 1997. Remote sensing in the surf zone–nontraditional methods. Pp. 91-95 in SACLANTCENTER Conference Proceedings Series CP-44, Lerici, Italy.

Kass, R.E., and L. Wasserman. 1996. The selection of prior distributions by formal rules. Journal of the American Statistical Association 91(435):1343-1370.

Latif, M., D. Anderson, T. Barnett, M. Cane, R. Kleeman, A. Leetmaa, J. O'Brien, A. Rosati, and E. Schneider. 1998. A review of the predictability and prediction of ENSO. Journal of Geophysical Research 103:14375-14393.

Lippmann, T.C., and R.A. Holman. 1989. Quantification of sand bar morphology: A video technique based on wave dissipation. Journal of Geophysical Research 94:995-1011.

Lippmann, T.C., and R.A. Holman. 1991. Phase speed and angle of breaking waves measured with video techniques, in coastal sediments, N. Kraus (ed.). ASCE, New York. Pp.542-556.

Mason, S.J., and N.E. Graham. 1999. Conditional probabilities, relative operating characteristics and relative operating levels, Weather and Forecasting American Meteorological Society. 14:713-725.

National Research Council. 1991. Symposium on Tactical Oceanography. National Academy Press, Washington, D.C.

National Research Council. 1992. Symposium on Naval Warfare and Coastal Oceanography. National Academy Press, Washington, D.C.

National Research Council. 1994. Proceedings of the Symposium on Coastal Oceanography and Littoral Warfare. National Academy Press, Washington, D.C.

National Research Council. 1996a. Expanding the Uses of Naval Ocean Science and Technology. National Academy Press, Washington, D.C.

National Research Council. 1996b. Proceedings of the Symposium on Tactical Meteorology and Oceanography: Support for Strike Warfare and Ship Self Defense. National Academy Press, Washington, D.C.

National Research Council. 1997a. Technology for the United States Navy and Marine Corps, 2000-2035: Becoming a 21st-Century Force, Volume 9, Modeling and Simulation. National Academy Press, Washington, D.C.

National Research Council. 1997b. Oceanography and Naval Special Warfare: Opportunities and Challenges. National Academy Press, Washington, D.C.

National Research Council. 1998. Improving Fish Stock Assessments. National Academy Press, Washington, D.C.

National Research Council. 2000a. Network-Centric Naval Forces: A Transition Strategy for Enhancing Operational Capabilities. National Academy Press, Washington, D.C.

National Research Council. 2000b. Oceanography and Mine Warfare. National Academy Press, Washington, D.C.

Nelson, S.B. 1990. Naval oceanography: A look back. Oceanus, 33(4):11-19.

Poulquen, E., A.D., Kirwan, and R.T. Pearson. 1997. Rapid Environmental Assessment, SACLANTCENTER Conference Proceedings Series CP-44, Lerici, Italy, pp. 290. 1997. Project (OCCAM). 1999. Journal of Geophysical Research 104:18281-18299.

Punt, A.E., and R. Hilborn. 1997. Fisheries stock assessment and decision analysis: The Bayesian approach. Reviews in Fish Biology and Fisheries 7:1-29.

Robinson, A.R., H.G. Arango, A.J. Miller, A. Warn-Varnas, P.M. Poulain, and W.G. Leslie. 1996. Real-time operational forecasting on shipboard of the Iceland-Faeroe frontal variability. Bulletin of the American Meteorological Society 72(2):243-259.

Saunders, P.M., A. C. Coward, and B. A. de Cuevas. 1999. Circulation of the Pacific Ocean seen in a Global Ocean Model. Ocean Circulation and Climate Advanced Modeling (OCCAM), Journal of Geophysical Research, in press.

Stockdale, T.N., A.J. Busalacchi, D.E. Harrison, and R. Seager. 1998. Ocean modeling for ENSO, Journal of Geophysical Research 103:14325-14355.

Stockdon, H.F., and R.A. Holman. 2000. Estimation of wave phase speed and nearshore bathymetry from video imagery. Journal of Geophysical Research 105:22015-22033.

Stommel, H. 1963. Varieties of oceanographic experience. Science 139:572-576.

Webb, D.J., B.A. de Cuevas, and A.C. Coward. 1998. The first main run of the OCCAM global ocean model. Southhampton Oceanography Centre, Internal Document No. 34 (available from: http://www.soc.soton.ac.uk/JRD/OCCAM/occam-papers.html). Pp. 43.

Appendixes

Appendix A

Committee and Staff Biographies

Rear Admiral Paul E. Tobin (U.S. Navy, ret.) graduated from the U.S. Naval Academy (1963) and received his M.S. degree in computer systems management from the U.S. Naval Postgraduate School in 1970. His career in the Navy spanned 35 years, during which he commanded the USS TATTNAL (DDG-19), the USS FOX (CG-33), and the Surface Warfare Officer School. RADM Tobin's final posting was as the Oceanographer of the Navy. His professional interests include shipboard engineering, computer systems, training and education, and oceanography. After retiring in 1998, RADM Tobin became the Executive Director of the Armed Forces Communications and Electronics Association and is currently a member of the Ocean Studies Board.

Thomas P. Ackerman received his Ph.D. in atmospheric science from the University of Washington (1976). Currently, he is a Battelle Fellow at the Battelle Pacific Northwest National Laboratory. His research interests are atmospheric radioactive transfer, planetary radiation budget, climatic effects of clouds and aerosols, aircraft and satellite observations of radiation fields, and ground-based remote sensing of cloud properties. Dr. Ackerman served on the NRC Advisory Panel for the International Satellite Climatology Project.

Arthur B. Baggeroer received his Sc.D. from the Massachusetts Institute of Technology (1968). He is presently a professor in MIT's Department of Ocean and Electrical Engineering. Dr. Baggeroer's research interests include sonar signal processing as applied to oceanographic research. A member of NAE, he has served on a number of NRC boards and committees, including the Ocean

Studies Board, the Naval Studies Board, and the Committee on Mine Warfare Assessment.

E. Ann Berman received her Ph.D. in Meteorology from the University of Wisconsin (1974). Dr. Berman supports the Navy in the area of operational concept development for emerging remote sensing systems. Currently, Dr. Berman is the President of Tri-Space, Inc., a remote sensing and software engineering company serving a broad range of environmental and security interests. Formerly, Dr. Berman served on several NRC committees, including the Board on Atmospheric Sciences and Climate and the Steering Committee on Improving the Differential Global Positioning System Infrastructure for Earth and Atmospheric Science Applications.

Stephen K. Boss received his Ph.D. in Marine Sciences from the University of North Carolina, Chapel Hill (1994). Currently, he is an Associate Professor in the Department of Geosciences and Director of the Environmental Dynamics Program at the University of Arkansas. His research interests are in the application of high-resolution geophysical methods to document and interpret the depositional geometry, stratigraphy, and regional geological history of sedimentary basins, continental margins, carbonate platforms, and lakes. Dr. Boss attended the Steering Committee for the Fifth and Sixth Symposiums on Tactical Oceanography and acted as a reviewer for both reports.

Tony F. Clark received his Ph.D. from the University of North Carolina (1974). He has been a professor in the Marine, Earth, and Atmospheric Sciences Department at North Carolina State University since 1996. Dr. Clark's research interests include underwater acoustics, oceanography, and marine geology and geophysics. A former naval officer with experience in submarine warfare and naval oceanography, he served on the NRC Steering Committee for the Symposium on Oceanography and Mine Warfare.

Peter C. Cornillon received his Ph.D. in Experimental High Energy Physics from Cornell University (1973). Currently, Dr. Cornillon is a Professor of Oceanography at the University of Rhode Island. Dr. Cornillon's current research interests include mesoscale dynamics of the upper ocean. Using satellite-derived sea surface temperature (SST), wind vector, and sea surface heights (SSH) fields in conjunction with in situ data and numerical model results, he is investigating Gulf Stream dynamics, Rossby wave forcing and propagation, and the characterization of SST fronts on regional (western North Atlantic) and global scales. Dr. Cornillon also teaches a graduate course on geophysical fluid dynamics.

Carl A. Friehe received his Ph.D. from Stanford University (1968). Currently, Dr. Friehe is a professor at the Atmospheric Turbulence Laboratory at the University of California, Irvine. Dr. Friehe's research interests include turbulence in the atmosphere, particularly that responsible for energy exchanges between the earth's land and ocean surfaces and the overlying atmosphere. He was a member of the Ocean Studies Board and served on several NRC committees, including the Marine Meteorology Study Panel, the Panel on Coastal Meteorology, and the Panel on the NOAA Coastal Ocean Program.

Eileen E. Hofmann received her Ph.D. in Marine Sciences and Engineering from North Carolina State University (1980). Currently, Dr. Hofmann is a Professor at the Center for Coastal Physical Oceanography at Old Dominion University. Dr. Hofmann's research interests include mathematical modeling of physical-biological interactions in marine food webs. Dr. Hoffman was a member of the Ocean Studies Board and has served on several NRC committees, including the Committee on Coastal Oceans and the Committee on Ecosystem Management for Sustainable Marine Fisheries.

Robert A. Holman received his Ph.D. in Physical Oceanography from Dalhousie University (1979). Currently, he is a Professor in the College of Oceanic and Atmospheric Sciences at Oregon State University. Dr. Holman's interests include but are not limited to measurement of nearshore waves and currents, application of remote sensing to nearshore processes and large-scale coastal behavior. Formerly, Dr. Holman served on several NRC committees, including the Steering Committee for the Sixth Symposium on Tactical Oceanography and the Steering Committee for the Symposium on Oceanography and Naval Special Warfare.

Gail C. Kineke received her Ph.D. from the University of Washington (1993). Currently, Dr. Kineke is an Assistant Professor in the Department of Geology and Geophysics at Boston College. Dr. Kineke's research interests pertain to sediment transport in coastal environments and are aimed at understanding how physical processes associated with rivers, waves, tides, and currents move sediment, transform the coasts, and deposit sediment in the marine environment.

John M. Ruddy received his Ph.D. in electrophysics from Polytechnic University (1967). Currently, Dr. Ruddy is the Technical Director and Deputy for System Development of the Ground-Based Missile Defense Program of the Missile Defense Agency, Department of Defense, Washington, D.C. Formerly, Dr. Ruddy was Vice President for the MITRE Corporation.

STAFF

Dan Walker is a senior program officer with the Ocean Studies Board, where he has been since July 1995. Since 1999, Dr. Walker has held a joint appointment as a Guest Investigator at the Marine Policy Center of the Woods Hole Oceanographic Institution. He received his Ph.D. in Geology from the University of Tennessee in 1990. Since joining the Ocean Studies Board, he directed a number of studies including *Oil in the Sea III: Inputs, Fates and Effects* (2002), *Clean Coastal Waters: Understanding and Reducing the Effects of Nutrient Pollution* (2000), *Science for Decisionmaking: Coastal and Marine Geology at the U.S. Geological Survey* (1999), *Global Ocean Sciences: Toward an Integrated Approach* (1998), and *The Global Ocean Observing System: Users, Benefits, and Priorities* (1997). A former member of both the Kentucky and North Carolina State geological surveys, Dr. Walker's interests focus on the value of environmental information for policymaking at local, state, and national levels.

John Dandelski is a Research Associate with the Ocean Studies Board and received his M.A. in marine Affairs and policy from the Rosenstiel School of Marine and Atmospheric Science, University of Miami. His research focused on commercial fisheries' impacts to the benthic communities of Biscayne Bay. As a graduate research intern at the Congressional Research Service he worked on fisheries and ocean health issues. Mr. Dandelski served as the RSMAS Assistant Diving Safety Officer and was involved in fisheries, coral, underwater archaeology, and ocean exploration projects.

Denise Greene has eight years of experience working for the National Academies and is currently a senior project assistant for the Ocean Studies Board.

Appendix B

The Role of Environmental Information in Naval Warfare

INTRODUCTION

As discussed throughout this report, the use of environmental information by U.S. Naval Forces evolved dramatically during the past 100 years. The rapid evolution in observation and prediction capabilities has reinforced the rapidly expanding need for one infrastructure driven by evolving naval tactics and weapons systems. Naval operations represent a complex interplay of a variety of missions; thus the meteorological and oceanographic (METOC) system must be flexible enough to meet a variety of demands on many timescales, driven by the specific objective of a given operation.

The various actions undertaken by different components of the fleet are grouped into mission warfare areas. These areas generally center on actions or assets with a common theme and include, among others, aviation and strike warfare, submarine and antisubmarine warfare, surface warfare, naval special warfare, and amphibious warfare (see Table 2-1).

Targeting Information and the Weather

The need for accurate targeting information is as important as being able to hit the target once it has been identified. Fixed targets can be located ahead of time by using satellite imagery or manned and unmanned reconnaissance aircraft. Again, the need for advanced weather information in the theater area is clear. Furthermore, it will be important to be able to locate enemy forces in real time. To best manage resources and plan tactics it will be necessary to be able to more accurately predict and identify current weather conditions in the search areas.

Battle Damage Assessment

Finally, it is important to perform battle damage assessment on targets that have been attacked. The scheduling of reconnaissance flights or satellite imagery to provide this postattack assessment will also be dependent on knowledge of current and predicted weather conditions in theater.

ENVIRONMENTAL INFLUENCES ON AVIATION AND STRIKE WARFARE

Aviation and strike warfare depend critically on accurate environmental assessments and forecasts, from winds for carrier operations to humidity, clouds, and haze effects on electrooptical sensor performance.

Carrier Operations

Carrier operations require a minimum of wind across the deck for safe take-offs and landings. Deck-level wind is the combination of atmospheric wind and the ship's forward velocity. Continuous in situ measurement of wind speed and direction is required for flight deck operations. Additionally, accurate forecasting of deck-level wind speed and direction is essential for vectoring the carrier, especially in geographic regions where maneuverability is constrained.

Radar

Many systems are directly affected by the atmospheric environment. Anomalous radar propagation due to humidity-dependent refractive index effects was noticed early in World War II. Radar range was highly variable, sometimes reaching over the horizon, due to refractive "ducting." Although the dependence of radar refractivity on air temperature, humidity, and pressure is known exactly, accurate prediction of these parameters is a continuing goal.

Optical Sensors and Lasers

Conditions common in the marine atmosphere limit the effectiveness of laser target designating systems (on both the laser designator and the optical sensor that acquires the laser beam energy) through refraction and scintillation ("twinkling") of the light beams. Obviously, fog, clouds, and marine aerosols are obscurants, and their prediction at the target is desirable. As naval operations focus on coastal regions, smoke and dust can also have a detrimental impact on the successful use of laser designators and optical sensors.

Present-Day At-Sea Data Collection

Good environmental information depends on good measurements and good forecasts. Typically, naval operations are at a disadvantage compared to land operations because of the lack of observations and the remoteness of operating areas. Ships give routine weather reports at synoptic times, but the instruments and information transmitted have not changed much since the sailing age. Ship-relative wind is obtained from a mechanical anemometer; true wind is calculated via a nomograph slide rule and logs of the ship's speed and heading. Temperature and humidity are measured with handheld wet-and-dry bulb psychrometers and bulk sea temperature from cooling intake seawater or a bucket sample. Pressure is from aneroid mechanical barometers in the wheel house, which are ported and manifolded to the outside (port and starboard). Sometimes aerographer's estimates of wind speed are obtained from the Beaufort scale of sea state appearance, which admittedly is judgmental but avoids distortion of the wind by the ship's structure. Wave data are by visual estimates. In contrast to the in situ measurements, frequent detailed satellite pictures are available to assist the METOC officer in evaluating the environmental scene.

Forecasting the weather has evolved considerably from Crick's case study method of comparing synoptic charts, Bjerkne's discovery of fronts, and Rossby waves to numerical integration of the governing equations on supercomputers. Numerical forecasting is, of course, the method used at the Fleet Numerical Meteorology and Oceanography Command (FNMOC) and has improved steadily. There is still a "person in the loop" in monitoring the model outputs, and the models are semiempirical and contain a wealth of experience from analysis of previous forecasts, data analysis, and results of focused experiments. The models rely on primitive shipboard data for initialization. Because the governing equations are extremely complex, they cannot be solved exactly, and errors due to inexact physical assumptions and/or inaccurate initial data propagate and can grow to unacceptable levels with increased forecasting time.

Key Problems

The environmental issues relevant to aviation and strike warfare cover a wide range. For situ ship data, the basic instrumentation on most ships is archaic. More complete systems for civilian use can be bought for less than $1,000. There has been no systematic study of the effect of the accuracy of routine ship observations on resulting forecasts. Weather centers such as FNMOC toss out "erroneous" data by a filter/comparison technique, thus pointing out the obvious distrust of the data and communications scheme. However, history is rife with examples of bad forecasts due to suppression of data simply because they did not fit a normal distribution. The proposed SMOOS upgrade to ship instrumentation is an improvement but does not address basic issues such as sensor location.

While the actual relative wind across the deck of a carrier is important for flight operations, wind and other measurements away from the flow blockage of the ship are required for numerical model input.

Smart Weapons Systems

Precision-guided munitions (PGMs) or "smart weapons" have an expanded role in modern warfare, especially in the past decade. Their value in warfare became evident during Desert Storm even though they accounted for only a small portion (approximately 7 to 9 percent) of the ordnance expended. This view was furthered in the Bosnian and Kosovo conflicts because of the increased percentage (70 percent of ordnance) of smart weapons used. These new weapons have been lauded for their accuracy, defined as a low circular error probability (CEP). Table B-1 shows the dramatic decrease in CEP in the past 60 years.

This improvement in performance has increased our ability to damage the adversary's equipment and warmaking capability while decreasing casualties on both sides. Because precision bombing can result in a significant reduction in civilian collateral damage and casualties (compared to previous wars), their use relative to conventional or dumb bombs has expanded significantly and can be expected to continue to grow. However, the very systems that make these weapons more precise also make them, in some instances, more sensitive to environmental conditions.

Development of Laser- and Operator-Guided and Terrain Mapping Weapons Systems

The first techniques (1976) made use of laser designators to illuminate the target and a laser seeker in the bomb to track and guide the weapon. Another approach (1985) uses TV (visible or infrared) to send signals back to a weapons

TABLE B-1 Improvement in Accuracy of Air-Delivered Weapons Since World War II

War	Weapon	CEP (m)
World War II	Gravity bomb	1,000
Korea	Gravity bomb	300
Vietnam	Gravity bomb	100
Desert Storm	Laser-guided bomb	8
	Tomahawk Block II	10
Bosnia	Laser-guided bomb	8
	Tomahawk Block III	3

system operator to remotely guide the bomb. Cruise missiles such as the Tomahawk use terrain-matching techniques en route to the target and digital optical matching for final target identification and acquisition.

Current and Next-Generation PGMs

The performance of many PGMs is highly dependent on weather factors, such as cloud cover, rain, smoke, and wind effects. The current generation of guided bombs and cruise missiles still depends on either an ability to visualize the scenes on the way to the target, the target itself, or both. New generations of guided bombs, such as the Joint Direct Attack Munition (JDAM) and cruise missiles such as the Tomahawk IV, will use the Global Positioning System (GPS) as well as inertial measurement units to provide higher accuracy in adverse weather conditions. The use of GPS can bring the CEPs down to 13 m in adverse weather conditions for either JDAM or the Tomahawk Block IV weapons. However, the ability to obtain CEPs of 3 m or less will still rely on favorable local weather conditions in the target area and electrooptical visualization methods upon reaching the target.

Optimizing the Performance of Weapons Systems

In order for the Navy (and land forces) in theater to be able to optimize the performance of their weapons systems, it will be necessary to (1) make use of all relevant weather data without regard to source and (2) be able to generate a continuous picture of weather in the theater in terms that are relevant to the warfighters and their smart weapons. This weather picture should be based on the same data and shared in a common format to all forces in the theater.

Use of Other (non-Navy) Weather Information

The Navy should be using data and contributing data to databases that are being used by the other services. Clearly each service will have its own way of manipulating and mapping data to support its own needs and weaponry. But insofar as they are operating in a common environment with similar or the same weapons, decisions and performance expectations should be based on the same weather picture.

Weather Data Dissemination

Data and data analyses should be available to all users via the Internet. This can be accomplished through ground-based or wireless networks depending on location and platform. The utilization of widespread and rapidly evolving commercially developed and supported infrastructures and infrastructure technologies should be exploited to the greatest extent possible commensurate with security needs. Consideration should be given to common networks and enterprise-level database and storage management technologies that are shared by the services. Each service or user can tailor products to its specific needs, but all of the most current data are made available to all.

Environmental Influences on Antisubmarine Warfare Systems

Acoustic Propagation

Because acoustic propagation, especially at low frequencies, is dictated by the nature of the physical environment, knowledge of the environment is fundamental to the performance of antisubmarine warfare (ASW) systems. Much of the existing knowledge is quite mature and is incorporated into the training of Navy sonar operators.

A number of texts are available that summarize this knowledge. Consequently, this discussion focuses on topics where environmental knowledge, or lack thereof, has impacted operational sonars and the development of more advanced systems. Furthermore, the environmental influence can be subdivided according to the surface water column (or volume) and the seafloor.

Acoustic Propagation at the Surface

Sonar performance is impacted by the sea surface, which is moving and can be rough primarily due to wind. Surface motion imparts Doppler shifts to a signal. An active system generates long narrowband signals to move a Doppler-shifted signal out of the clutter. The surface imparts a Doppler spreading that degrades performance. Doppler effects are particularly important for fixed systems where platform motion is mostly eliminated and Doppler is the primary observable parameter.

Surface roughness becomes important when the Rayleigh number, which scales according to the RMS (roughness to wavelength), becomes high. At this point the surface can no longer be treated as a simple reflector. The roughness leads to angular spreading and ray/mode coherence. This angular spreading can be measured by the large-aperture systems now in use. The ray/mode coherence is a fundamental issue in ranging systems that exploit vertical multipath. Again, these issues are more important for fixed systems. Additionally, surface roughness impacts the performance of wake detection systems. The same surface processes, such as wave breaking and wind/wave interaction, are also important sources of ambient noise.

Bubble processes are a very important component of the surface. Very small fractions of bubbles lead to significant changes in sound speed, which modulates the reflection/refraction properties of sound near the surface. Moreover, bubbles are both efficient scatterers and sources of ambient noise.

While there is still much to do with regard to modeling acoustic interaction with the surface, environmental drivers of surface effects according to region and time of year are important predictors of modeling sonar performance.

Acoustic Propagation in the Water Column

The sound speed profile (SSP) of the water column is one of the fundamental quantities needed for predicting propagation of an acoustic signal. All the basic features associated with this are well known and are tabulated in operational databases. There are, however, significant gaps in these databases for operationally important regions.

Long-range detection systems of Cold War vintage exploited SSP data extensively. Since these were primarily low-frequency, deep water, low-resolution systems, the databases were usually adequate for coarse predictions. Nevertheless, internal waves that led to scattering could be a factor. Over the years important processes such as seasonal surface duct formation and focusing effects were understood. However, environmental questions do remain for long-range systems. These include the impact of internal waves and mesoscale eddies on the coherence of multipath/signals. Specification of fronts and eddies is also important for long-range predictions.

Sonar detection ranges are now much shorter because of advances in reducing ship noise (referred to as "quieting"). Thus, ship detection is generally regarded as possible at distances of tens versus hundreds of kilometers. SSPs remain important because of the significant role they play in ranging algorithms used to interpret acoustic signals. While these can usually be measured in situ, it is still important to know the general characteristics of an area for a specific time of year.

Probably, the most important volumetric aspect concerns littoral waters. First, they are very dynamic with short scales. A range curtain governed by interaction of the SSP and the geoacoustics of the seabed effectively limits sound propagation ranges. Moreover, many of the processes, including internal waves, soliton generation, propagation across slopes, and regions of rapidly changing bathymetry, require better environmental information than is currently available.

Acoustic Propagation at the Seabed

One of the most important aspects of acoustic propagation is whether a path is bottom interacting. The seabed is a loss mechanism that can, during reflection, change the frequency of and/or attenuate the emitted sound wave. Current knowledge of the bottom at specific locations is inadequate for state-of-the-art predictions. The models used lead to prediction ranges that can vary by a factor of 3. While we have detailed charts of the bathymetry of the deep ocean, such knowledge for important littoral areas is lacking. Moreover, at low to medium frequencies, where there is significant geoacoustic interaction, the models and database simply do not give useful predictions. Many of the data are gathered by normal-incidence, high-frequency probing, while the concern is with low grazing angle, low-frequency regimes. Transmission loss measurements are made; however, the data reduction often fails to capture important features and it is next to impossible to get at the original data.

Geoacoustic data are absolutely necessary for predictions in littoral environments because acoustic interaction with the bottom is unavoidable. Detailed bathymetry and subseabed geoacoustics are imperative. Roughness and coherence scales are needed for input to predictive models. Understanding of the controls that the environment places on acoustic properties is certainly one of biggest factors now missing.

Effects of Active Acoustical Transmissions on Marine Mammals

Another area in ASW where environmental data and support are important is related to the effects of active acoustic transmissions on marine mammals. This is a highly emotional issue driven in most cases not by science but by the media and nongovernmental organizations' recent outcry over the potential for harm to marine mammals by active sonar transmissions. The intensity of this opposition is highlighted by a recent lawsuit by the National Resources Defense Council and several other nongovernmental organizations against the Littoral Warfare Advanced Development program at the Office of Naval Research. It is also significant that the SURTASS-LFA program at SPAWAR has been prohibited from conducting sea tests for the past six years because of public opposition to sonar transmissions in the oceans. Environmental activists have also targeted active operational sonars as being deleterious to marine mammals.

More studies and data are needed to assist the Navy in understanding and explaining the effects of active transmissions on marine mammals. The effect of active transmissions (as a function of frequency and source level) on marine mammal physiology, migration patterns, feeding habits, and general well-being must be determined. As a part of this requirement, environmental data are also needed to facilitate detection of marine mammals in areas where active sonars are to be used. Passive and active scenarios are both used to detect marine mammals. Environmental data to support these efforts consist of basic acoustic parameters such as propagation loss, ambient noise detection threshold, and source level.

Environmental Influences on Naval Special Warfare and Amphibious Warfare

Amphibious warfare and naval special warfare (NSW) are the areas of naval warfare most prominently involved in recent shifts in emphasis on shallow water (or "brown water") naval operations. Both areas focus activities from shallow waters through the beach to inland areas, both operate at short time- and space scales, and both are heavily affected by many aspects of the coastal environment. While they can share a common battlespace, differences in their mission focus and style suggest separate discussion of each problem and their METOC sensitivities. In particular, NSW emphasizes clandestine operations by small groups to achieve specific limited objectives while amphibious warfare operations typically involve much extensive and more overt troop activities.

Naval Special Warfare (NSW) and METOC

Since their creation in 1970, the SEALs of NSW have developed and maintained a reputation as an elite and effective force for special operations. Typically working in teams of 16 members (platoon) or squads of 8, their missions are usually characterized by the need to operate stealthily and remain undetected from insertion through extraction. When compromised, SEAL units are capable of demonstrating a show of force out of proportion to their small numbers through intense, often unexpected, violence of action. Swimmer operations figure strongly in the mission profile, and the METOC role is of critical importance in forecasting environmental conditions necessary for planning.

In 1997 a tactical oceanography symposium initiated a dialogue between academic oceanographers and NSW operators. Much of the material from this section is extracted from the report of that meeting (National Research Council, 1997).

NSW Mission

The small, highly trained nature of NSW forces is conducive to a range of missions. These include counterterrorism, counterproliferation, special reconnaissance, psychological operations, direct action, foreign internal defense, civil affairs, information warfare, and unconventional warfare. In addition, NSW is involved in a range of collateral activities, including combat search and rescue, counter-drug, counter-mine, and humanitarian assistance.

Mission Planning

Although the specifics of mission activities vary enormously, they all involve the five phases of insertion, infiltration, actions at the objective, exfiltration, and extraction. Mission planning occurs on a 96-hour countdown to operation. METOC information most strongly influences infiltration and exfiltration.

Amphibious Warfare

Amphibious operations involve assault from the sea of an enemy-held coastal region and securing the subsequent Logistics-Over-The-Shore capability for resupply of ongoing operations. While the element of surprise is important, operations are not usually clandestine and often involve combat. There is a necessity for speed and particularly in-stride mine-sweeping capability. However, troop sizes generally include thousands of Marines, and transit is by armored and substantial vehicles rather than the swimmer operations of the SEALs. Thus, environmental factors such as water temperature are not as directly important while other factors such as visibility, wave height, and electromagnetic ducting may be more important.

Mission Planning

Mission planning cycles for amphibious operations can be substantially longer than the 96-hour SEAL time line due to the greater logistical complexities required. However, METOC information can clearly stop a mission, and up-to-date METOC information is required right up to the final go/no go decision.

Role of the Environment

The required resolution of environmental knowledge in the littorals is substantially finer than for traditional open-ocean purposes and increases with proximity to the beach. At the extreme, for example, the surf zone contains rip channels that may offer safe gaps for landing vehicles but deadly countercurrents for swimmers. These features are only tens of meters wide but are sufficiently stable to make 24-hour reconnaissance data useful for planning.

Because of the required fine granularity of predictions and the generally changeable nature of the nearshore environment, METOC information must be current and of high resolution. To achieve this purpose requires quantitative use of remote sensing data (ground truthed with some in situ data, if possible) fused with numerical models for complete battlespace characterization.

METOC Variables for Nearshore Operations

Discussions at the 1997 tactical oceanography symposium led to the development of a full list of important METOC variables for nearshore operations, their level and nature of impact, and the state of scientific understanding or capability for their estimation. Major environmental factors include:

- Bathymetry—of pervasive importance with either direct impact (obstacle) or indirect impact (e.g., cause of wave shoaling and breaking). Shelf life of data is short in surf zone, longer offshore.

- Waves and surf—major impact on all operations as described by operational limits. Refraction can lead to substantial along-coast variability that can be used to an advantage.
- Currents—major impact on swimmer operations, less so for amphibious vehicles. Can be modeled, given reasonable bathymetry data.
- Tides—affects clearance over shoals and obstacles, workable depth. Tidal currents can be strong in some sites.
- Water temperature—major impact on swimmer operations. Can be predicted or remotely sensed.
- Turbidity—useful for clandestine swimmer operations but can hinder navigation or obstacle (and mine) search.
- Bioluminescence—major impact on clandestine SEAL operations.
- Wind—direct effect on parachute operations, indirect effect through waves.
- Precipitation—visibility effect (potentially positive) with other impacts on fatigue and discomfort.
- Atmospheric visibility—effect can be positive or negative.
- Humidity—affects electromagnetic ducting and fatigue in warm climates.

Environmental Influences on Mine Warfare and Mine Countermeasures

Since the refinement of sea mines and sea-mining strategies during World War I, naval mine warfare has figured prominently in all major conflicts involving U.S. Naval Forces. Indeed, since the Korean War (1950-1953), sea mines have been responsible for more U.S. ship casualties than all other forms of attack combined (Avery, 1998).

Mine Warfare

The importance of mine warfare (MIW) in both offensive and defensive operations lies in the efficiency of mines as force multipliers:

- Mines are inexpensive and can be technologically simple to build; the simplest forms of sea mines rely on 19th-century technology that is still effective (National Research Council, 2000b).
- Mines can require very little maintenance. Once deployed in the marine environment, mines can persist as active threats for years.
- Mines are easily deployed from almost any platform. Swimmers or divers, small boats, large ships, submarines, or aircraft are capable of mine-laying operations.
- Mines are small and stealthy; the small size of mines combined with their frequent deployment in shallow water marine environments makes their detection and neutralization difficult.

Technologies for deploying, concealing, and activating mines are becoming increasingly sophisticated. Mine inventories of a number of nations include traditional bottom, moored, and drifting mines that are activated by contact with a target as well as modern designs that avoid detection using unconventional shapes and anechoic coatings. In addition, many modern mines are activated through influence (magnetic, acoustic, or pressure) from the target or via remote control.

MIW Battlespace

The MIW battlespace ranges from the littoral zone (coastal rivers and estuaries, beaches, and surf zone) across very shallow water (10 to 40 feet) and shallow water (40 to 200 feet) environments to deep water (>200 feet) settings offshore continental margins, island margins, and narrow seaways. Environmental parameters critical to MIW and mine countermeasure (MCM) operations vary among these zones (Table B-2) but generally become more complex and dynamic with decreasing water depth and proximity to shore.

TABLE B-2 Most Common Oceanographic Parameters Used for Nearshore MIW Operations and Their Importance in Different Water Depth Zones as Defined by the Navy

	Riverine/ Estuarine[a]	Surf Zone	Very Shallow Water	Shallow Water	Deep Water
Seafloor					
Bathymetry	H	H	H	H	M
Sediment grain size	H	M	H	H	M
Seafloor clutter density	M	M	H	H	L
Bottom roughness	M	M	H	H	L
Mine burial	H	H	H	H	L
Water column					
Currents	H	L/H[b]	H	H	M
Water clarity	M	L	H	H	M
Temperature and salinity	H	L	M	H	H
Waves	L	H	H	M	L
Acoustic properties	H	N/A	H	H	H

[a]Riverine and estuarine environments pose unique problems not addressed in National Research Council (2000).

[b]For the surf zone, information on wind-generated currents is of low priority as these currents are generally overwhelmed by wave-generated currents.

H = high, a parameter that is essential for MIW operations in this depth zone.
M = medium, a parameter that is useful for MIW operations in this depth zone.
L = low, a parameter that is of little use for MIW operations in this depth zone.

SOURCE: National Research Council (2000b).

APPENDIX B

Deployment of mines of all types in these zones greatly impairs surface and submarine warfare operations, requiring a significant dedication of assets for mine detection, mine sweeping, and mine neutralization operations. Such operations impede the progress of offensive naval operations, degrade or completely remove the element of surprise, and potentially expose U.S. combat forces to hostile activity for longer periods of time. Thus, it is imperative that efficient MCM are developed and employed to remove mine threats in advance of U.S. surface and submarine naval operations. Understanding the dynamic oceanographic conditions within each MIW zone is an important component of evolving MCM strategies and techniques.

Oceanographic Parameters for Mine Deployment

Environmental Parameters of Importance to MIW and MCM Operations

Environmental parameters of importance to MIW and MCM operations are divided among two general categories: those related to the seafloor and those processes and phenomena related to the overlying water column. Each of the parameters identified in Table B-1 is discussed briefly below.

Seafloor Processes and Phenomena

Accurate bathymetry is useful and in some instances essential across all areas of MIW operation. Areas of complex bathymetry make mine detection difficult. Associated with complex bathymetry, the seafloor may display extreme roughness and clutter density, making detection by standard acoustic methods difficult or impossible. Mines (especially bottom mines with nontraditional shapes) may be concealed by irregular seafloor relief or by acoustic shadows created by seafloor relief elements in rocky areas. These problems are particularly acute in the surf, very shallow water, and shallow water zones. In the riverine/estuarine environments, surf zone and very shallow water areas characterized by unconsolidated sediments and vigorous physical processes (e.g., wind waves, wave-generated currents, tidal currents) redistribute sediments through erosion or deposition sufficient to significantly alter bathymetry. Thus, seafloor bathymetry is dynamic and can vary on very short timescales (as short as tidal half cycles). While complex bathymetry is less problematic in deep water, where moored or drifting mines are more commonly deployed, seafloor relief can influence MIW operations by affecting the choice of mooring platform or by limiting mooring locations.

Sediment grain size (texture) is important to MIW operations across all such zones. Very large sediment particles (e.g., boulders) impart extreme seafloor bottom roughness and clutter densities that make mine detection difficult, especially for bottom mines with nontraditional shapes. Also, large sediment particles can create acoustic or optical shadows that conceal smaller mines. Alternatively, fine-grained sediments typical of many estuarine and riverine environments pose problems for mine deployment (MIW operations) and mine detection (MCM operations). In response to spatial variability in salinity, fine-grained sediments may settle rapidly as aggregates, forming regions of very soft, unconsolidated sediments or fluid mud. These deposits can be centimeters or even meters thick along muddy coastlines and frequently produce "false bottom" echoes on traditional echo sounder systems, making accurate bathymetric mapping, bottom sediment characterization, and mine detection difficult. Mines may settle through mud and fluid mud deposits, becoming completely submerged and undetectable in the sediment but remain as easily encountered and effective threats to naval operations. Sediment textures are less problematic in the deep water zone where moored or drifting mines are more commonly deployed. However, seafloor textural attributes may influence the choice of mooring platform or mooring sites.

Clutter density and bottom roughness are closely related environmental parameters affecting MIW and MCM operations. In areas where the seafloor is characterized by a high degree of bottom roughness (e.g., glacial marine boulder fields, volcanic boulder fields, coral reefs), the output from acoustic or optical sensors is characterized by "clutter"—extreme scattering of sensor impulses such that the resultant imagery is irregular and difficult or impossible to classify. The poor quality of acoustic and optical imagery over seafloor with extreme bottom roughness makes mine detection very difficult or impossible, especially for bottom mines designed with nontraditional shapes or anechoic coatings.

Mine burial in the riverine/estuarine zone and the surf zone through shallow water environment is influenced by sediment grain size and the specific dynamics of fluid motion and sediment transport imparted by seafloor roughness and larger-scale bathymetric features (e.g., underwater headlands, shoals, reefs). In the riverine/estuarine zone through shallow water zone, mine burial interferes with MCM operations because buried mines are difficult to detect. In the four shallow water environments, buried mines remain effective and lethal despite burial, so their detection and neutralization are of paramount concern. In the deep water zone, mine burial is less common because moored or drifting mines are typically deployed. In addition, mine burial in the deep water zone is likely to result in reduced weapons effectiveness and is therefore of somewhat limited concern.

Water Column Processes and Phenomena

Currents (e.g., river currents, wave-generated currents, wind-generated currents, and tidal currents) are very important parameters for MIW activities. In the riverine/estuarine zone, currents may be dominantly unidirectional (related to river outflow) or strongly bidirectional (related to tidal oscillations) and may vary in strength on many temporal scales from hours to seasons. Currents in this zone are often sufficiently powerful to effect erosion, with resultant transport and redeposition of sediments and mine scour or burial. Opposing river outflow currents and tidal currents may result in very turbulent flow and mixing of water masses with quite different temperature and salinity characteristics, and these can impact the performance of acoustic and optical sensors used for mine detection.

In the surf zone, powerful currents generated by breaking waves can affect transport and redistribution of mines or transport large quantities of sediment, resulting in mine burial on a variety of timescales important to MIW operations. In addition, the surf zone is an environment where extreme turbulence and bubble saturation in the water column renders acoustic or optical detection systems useless. Thus, physical conditions in the surf zone water column continue to pose very challenging problems for MIW and MCM operations. In very shallow water and shallow water environments, wind-driven currents are somewhat more predictable than in the surf zone, but their impacts on MIW are no less profound. Strong currents cause significant scour around bottom mines or mine moorings and greatly affect the distribution or track of drifting mines. Alternatively, strong shelf currents or bottom-feeling waves can transport sediment, resulting in mine burial, which makes detection difficult.

Water clarity is inversely related to turbidity or opacity. Water clarity affects both MIW and MCM operations by affecting the performance of optical sensors (divers or other optical sensing devices) and their capability to detect mines or accurately map bottom features and bathymetry. In riverine/estuarine settings, water clarity can vary from clear to opaque, and vice versa, on timescales of less than an hour due to rapid resuspension and settling of sediments induced by tidal flow and water mass mixing. In the surf zone, water is typically rendered nearly opaque to optical sensors as a result of breaking waves and consequent bubble saturation and sediment suspension in the water column. In shallow and very shallow water zones, water clarity may vary seasonally according to variations in primary productivity, incursion/excursion of various water masses, and freshwater and sediment outflow from coastal rivers or estuaries. In deep water zones, optical properties of water are controlled primarily by abundance and type of plankton. Plankton abundance may vary diurnally to seasonally, with associated impacts on visual recognition of drifting or moored mines.

Temperature and salinity are critical environmental parameters affecting MIW and MCM activities in a variety of ways. Temperature and salinity variations may impart stratification to the water column with concomitant influences on the acoustic properties of the water column and the performance of acoustic sensors used for mine detection. Temperature and salinity variations in riverine/estuarine and coastal settings may result in density fronts in the water column that affect acoustic sensor performance. Suspended sediments may become trapped on these fronts, resulting in turbid horizons in the water column, or mixing of water with different salinities may induce flocculation of very fine-grained sediments, resulting in settling as loose aggregates on the seafloor as mud or fluid mud deposits. In shallow and deep water zones, oceanographic fronts and water column stratification related to temperature and salinity differences affect acoustic properties of the water column and thus the performance of acoustic sensors.

Waves are of critical concern in the surf zone through the shallow water zone because of their capacity to generate strong currents (e.g., in the surf zone), scour and redistribute bottom sediment (surf zone through shallow water), induce turbulent mixing in the water column that may affect both optical and acoustic properties of the water column, and affect nearshore MIW and MCM operations. In the surf zone, breaking waves render the environment opaque to acoustic sensors. Breaking waves generate strong longshore and rip currents capable of redistributing mines, creating scour around mines, or transporting sediment and burying mines on dynamic temporal scales. In shallow and very shallow water zones, bottom-feeling waves may induce scour around mines or affect sediment transport, resulting in mine burial. In deep water, white caps or rough sea surface conditions with large or confused wave patterns may make visible detection of drifting mines very difficult.

The acoustic environment of the ocean is very complex, especially in littoral settings where MIW activities are most prevalent. Nearshore environments (riverine/estuarine, surf zone, shallow water, and very shallow water) are characterized by high reverberation, high ambient noise, and both vertical and horizontal water column heterogeneities, all of which affect acoustic properties of the ocean environment. As such, any physical, chemical, biological, or geological phenomena that affect the water column have potential impacts on its acoustic properties and the performance of acoustic sensors used for mine detection.

Environmental Influences on Biochemical/Environmental Warfare

Biological Weapons

Biological weapons include bacteria, viruses, and toxins that are spread deliberately in the air, food, or water to cause disease or death to humans, animals, or plants. Examples include plague, smallpox, and anthrax. Biological agents tend to be persistent and can have a delayed effect extending for days or weeks. Biological weapons are attractive to so-called rogue states or nonstate actors such as terrorist groups because they provide a means of waging asymmetric warfare against an adversary with superior military capabilities. They are easy to acquire, since agents occur in nature as the causative agents of disease. Depending on the biological agent, small amounts are capable of creating significant disruption, fear, and a number of deaths.

Chemical Weapons

Chemical weapons would likely be dispersed in the air for the purpose of rapidly debilitating or destroying humans, animals, or plants. Examples include mustard gas, sarin, and napalm. Chemical agents are usually volatile (although some can be persistent). It takes a large volume of chemicals and thus a large number of aircraft to deliver chemical agents to their targets (i.e., to attack a city, seaport, or battle group). For this reason a chemical attack would likely be more obvious than a biological attack and easier to defend against.

Scenarios

One scenario might be a single biochemical attack against a single U.S. naval vessel or naval base that does not result in large casualties or affect the viability of military operations (USS Cole-type scenario). An opponent might consider such an attack as a propaganda medium and a way to demoralize U.S. military personnel.

A second scenario might be an aggressor who decides that chemical/biological warfare may diminish the willingness or capability of the United States and its allies to intervene. In this case, the aggressor might try to use such weapons to kill significant numbers of U.S. and allied military personnel and to raise the cost of defending against aggression well above what it would be otherwise.

Role of Environmental Information

The environment (air and water pathways) is the transmission medium for biochemical weapons of mass destruction. Environmental pathways should be addressed and modeled by the METOC community. Some examples are:

- A dispersion zone of influence forecast might be a worthwhile product for special forces and other forward-deployed forces.
- Mesoscale modeling and forecasting.
- Urban terrain and dispersion.
- Distributed sensors to feed mesoscale transport models.

METOC should take a role similar to the role meteorologists have taken for air pollution issues. This should include establishing siting criteria for biodetectors on ships and determining the type and number of sensors required to estimate the duration of a shipboard chemical attack.

Important Parameters in Air

Environmental parameters important for airborne dispersal of biological or chemical warfare agents are temperature and humidity; wind speed; wind direction; turbulence parameters for urban or rural environments; solar insolation, photooxidation, and/or decay; and precipitation and precipitation rate.

Important Parameters in Water

Environmental parameters important for dispersal of biological or chemical warfare agents in water are currents, temperature and salinity, waves, and turbidity.

Environmental Influences on Multi Mission Operations

Multi mission naval operations are those that incorporate elements of power projection, air-sea dominance, and deterrence of hostile actions into a complex 4-D battlespace, often to be coordinated with operations of other armed forces of the United States or coalition partners. By their very nature, multi mission operations conducted by naval forces will require maximum utilization of navy METOC assets in such a way that the full spectrum of naval METOC products will be required throughout the mission time line.

Multi Mission Scenario

A hypothetical multi mission scenario might include requirements for environmental information relevant to (1) naval special warfare activities, (2) suppression of enemy air defenses, (3) tactical strike warfare, (4) time-critical strike warfare, (5) MIW and MCMs, (6) naval amphibious warfare, (7) naval surface warfare (primarily ship self-defense), and (8) submarine and anti-submarine warfare. Clearly, requirements for environmental information in support of each of the operational elements of the multi mission scenario above are diverse. In addition, it will be necessary to disseminate relevant environmental information to a broad spectrum of fighting forces across a number of armed services. As such, interoperability and compatibility of environmental data or data products are overriding requirements for successful execution of multi mission objectives. A brief summary of the range of environmental information that might be required for successful conduct of multi mission operations is presented below. This summary is intended to serve as an illustration of the complexity of METOC requirements for this type of mission and is not necessarily a comprehensive review of all potential information needs. Instead, the review below is intended to stimulate exploration of strategies to achieve the desired sensor and information coverage within the multi mission battlespace necessary to meet the needs of the various warfighters.

Naval Special Warfare

Naval special warfare (NAVSPECWAR) operations typically involve covert deployment of SEAL teams in hostile territories. As the name implies, SEALs may be deployed from the sea (submarines, swimmer delivery vehicles, fast

TABLE B-3 General Summary of Environmental Data Needs for NAVSPECWAR Operations

Deployment Style	Environmental Information Needs
Sea	Ocean temperature, currents, tides, winds at sea, surf/sea state (period, height, wavelength, direction, steepness, "groupiness"), salinity, water optical properties, bioluminescence, dangerous marine organisms, ocean or other environmental toxins, near shore acoustics, bathymetry, lunar phase/illumination, biofouling, bottom character
Air	Surface winds, winds aloft (speed, direction), EM ducting (effects on radar propagation/detection and communications), temperature and humidity, temperature and humidity profiles with altitude, cloud cover, lunar phase/illumination, visibility, aerosols (type and quantity), EO properties of atmosphere, atmospheric boundary layer dynamics, terrain, vegetation, precipitation forecasts
Land (amphibious)	Winds at sea, surf/sea state (period, height, wavelength, direction, steepness, "groupiness"), bathymetry, beach trafficability, terrain, vegetation, EM-EO ducting, slant range visibility, aerosols, temperature, humidity, precipitation, lunar phase/illumination

SOURCE: National Research Council (1997).

surface vessels, swimming through open water), air (fast-rope helicopter insertion, parachute drops from any altitude), or land (amphibious landing). Environmental information needs for each of these deployment strategies are different, and a general summary is provided for each in Table B-3.

Suppression of Enemy Air Defenses

Suppression of enemy air defenses (SEAD) will typically be accomplished using various strike warfare tactics and weapons systems. Increasingly, there is a need for precise target identification and location for these missions, and often PGMs such as cruise missiles or other "smart" weapons (e.g., Joint Direct Attack Munition), radar-homing weapons (e.g., HARM), electrooptical weapons (e.g., Maverick), and various gliding ordnances will be utilized in addition to potential use of close air support aircraft (A-10, AV-8, A-6, F/A-18, AH-1F, AH-1W, AH-64, RAH-66). At least some of the weapons systems used for SEAD will depend on NAVSPECWAR operators in the theater providing targeting intelligence and laser designation of targets. Use of these weapons platforms for SEAD operations will require a variety of METOC forecast products as well as more specific environmental data needs (summarized in Table B-4). Again, the list presented in Table B-4 is not comprehensive but is intended to provide a sense of the diversity of environmental data needs for SEAD operations.

TABLE B-4 General Summary of Environmental Data Needs for SEAD Operations

SEAD Weapons Systems	Environmental Information Needs
Cruise missile	Digital terrain model, complete forecast along flight path and for duration of flight, winds aloft, winds at target, aerosols, slant range visibility, cloud cover, precipitation, precise target identification and location, space weather (if relying on direct GPS communication), vertical wind shear, atmospheric turbulence
EM-targeting weapons	EM ducting, winds aloft, winds at target, vertical wind shear, atmospheric turbulence, aerosols, precipitation, precise target identification and location
EO-targeting weapons	Atmospheric refraction, refractive index, visibility, aerosols, slant range visibility, winds aloft, winds-at-target, precipitation, cloud cover, precise target identification and location
Gliding ordnance	Precise target identification and location, winds aloft, winds at target, vertical wind shear, atmospheric turbulence, precipitation, space weather (if relying on direct GPS communication)
Close air support	Precise target identification and location, winds aloft, winds at target, vertical wind shear, atmospheric turbulence, precipitation, space weather (if relying on direct GPS communication), slant range visibility, lunar illumination, weather fronts and severe weather systems, cloud cover, cloud ceiling

SOURCE: National Research Council (1996b).

Tactical and Time-Critical Strike Warfare

Strike warfare involves the use of combat aircraft, cruise missiles, and (as recently demonstrated in Afghanistan) armed unmanned aerial vehicles (UAVs) to (1) penetrate an adversary's defenses from the air, (2) deliver precision ordnance on either fixed or mobile targets, and (3) assess battle damage while ensuring a safe return to the base of operations. In the context of naval strike warfare, this means air operations from aircraft carriers or ship-launched cruise missiles. Increasingly, there is recognition that many strike missions (especially those targeting mobile or opportunistic targets) have a time-critical dimension such that weapons systems must be selected and delivered precisely on target with a very short decision cycle. Such time-critical assaults require near realtime environmental information from battlefield sensors, and these data must be processed and disseminated from the sensors to the weapons system within minutes. Such rapid target identification, location, and destruction place extraordinary demands on METOC capabilities. Factors affecting tactical and time-critical strike missions are summarized in a general way in Table B-5. It is important to note that a significant proportion of naval strike missions continue to

TABLE B-5 General Summary of Environmental Data Needs for Tactical and Time-Critical Strike Operations

Tactical/Time-Critical Strike/ Warfare Systems	Environmental Information Needs
Combat aircraft, cruise missiles, unmanned aerial vehicles	Precise target identification and location. EM-EO ducting, winds aloft, winds-at-target, vertical wind shear, atmospheric turbulence, aerosols, humidity, precipitation, slant-range visibility, cloud cover, cloud ceiling, precipitation, weather fronts, severe weather, icing conditions aloft (for UAVs), space weather (if relying on direct GPS communication), atmospheric refraction effects, atmospheric scintillation, forecast along flight path and for duration of strike operations, precipitation rate, characteristics of surface and near-surface clutter

SOURCE: National Research Council (1996b).

be adversely affected by weather. (National Research Council, 1996b, reports that 90 percent of strike missions from 1992 to 1995 suffered from weather effects.)

Mine Warfare and Mine Countermeasures

Technologies for deploying, concealing, and activating mines are becoming increasingly sophisticated, and MIW and MCM in multimission scenarios greatly complicate efficient execution of these missions. In the worst cases mines may cause significant loss of naval assets (e.g., ships, submarines, amphibious assault craft) and personnel. In less severe cases, the presence of mines requires a significant dedication of assets for detection, sweeping, and neutralization while at the same time impeding progress of offensive operations, removing the element of surprise, and exposing U.S. combat forces to hostile action for longer periods of time.

MIW Battlespace

The MIW battlespace ranges from the littoral zone (coastal rivers and estuaries, beaches, and surf zone) to deep water settings offshore continental margins, island margins, and narrow seaways. Environmental parameters critical to MIW and MCM operations vary among these zones (see Table B-6) but are generally divided between processes and phenomena related to the seafloor and processes and phenomena related to the overlying water column. As a general rule, these processes become more complex and dynamic with decreasing water depth and

TABLE B-6 Most Common Oceanographic Parameters Used for Nearshore MIW Operations and Their Importance in Different Water Depth Zones as Defined by the Navy

	Riverine/ Estuarine[a]	Surf Zone	Very Shallow Water	Shallow Water	Deep Water
Seafloor					
Bathymetry	H	H	H	H	M
Sediment grain size	H	M	H	H	M
Seafloor clutter density	M	M	H	H	L
Bottom roughness	M	M	H	H	L
Mine burial	H	H	H	H	L
Water column					
Currents	H	L/H[b]	H	H	M
Water clarity	M	L	H	H	M
Temperature and salinity	H	L	M	H	H
Waves	L	H	H	M	L
Acoustic properties	H	N/A	H	H	H

[a]Riverine and estuarine environments pose unique problems not addressed in National Research Council (2000b).
[b]Within the surf zone, information on wind-generated currents is of low priority as these currents are generally overwhelmed by wave-generated currents.
H = High, a parameter that is essential for MIW in this depth zone.
M = Medium, a parameter that is useful for MIW in this depth zone.
L = Low, a parameter that is of little use for MIW in this depth zone.

SOURCE: National Research Council (2000b).

proximity to shore. An understanding of the dynamic oceanographic conditions in each MIW zone is an important component of evolving MCM strategies and techniques.

Naval Amphibious Warfare

Naval amphibious warfare involves assaults from the sea to land occupied by hostile forces, clearing land of those forces, securing a forward-operating base, and preserving capability to move additional warfighters and warfighting equipment across the beach. Amphibious warfare shares many similar environmental needs with NAVSPECWAR operations, though amphibious operations are typically overt rather than clandestine. Warfighting equipment and personnel are usually transported from sea to land in air cushion vehicles, amphibious landing craft, and amphibious armored vehicles. Oceanic, atmospheric, and geological

TABLE B-7 Environmental Parameters Impacting Amphibious Warfare

- Currents
- Tides
- Winds at sea
- Surf/sea state
    o Period
    o Height
    o Wavelength
    o Direction
    o Steepness
    o "Groupiness"
    o Bioluminescence
- Bathymetry
- Coastal geology
- Terrain
- Beach trafficability
- Vegetation
- Offshore obstructions (reefs, rocks, etc.)
- Lunar phase/illumination
- Cloud cover
- Cloud ceiling
- Fog
- Aerosols
- Horizontal visibility
- Radar-frequency propagation
- Atmospheric refraction
- EM-EO ducting
- Precipitation

features of the environment can affect the performance of these platforms. Some of the environmental parameters affecting amphibious operations are listed in Table B-7.

Surface Warfare (Ship Self-Defense)

Surface warfare may involve direct engagements against enemy vessels but more commonly will be focused on ship self-defense, especially in the littoral zone. Increasingly, ship self-defense is defense against airborne threats—either aircraft or missiles and sea-skimming missiles. As such, environmental parameters of importance to ship self-defense are very similar to those of importance to tactical and time-critical strike warfare. Indeed, ship self-defense might be reasonably categorized as a form of time-critical strike warfare because of the (usually) very short time to detect, acquire, and neutralize incoming sea-skimming missiles. Table B-8 provides an overview of environmental parameters important to ship self-defense.

TABLE B-8 Environmental Parameters Impacting Ship Self-Defense

- Winds at sea
- Surf/sea state
  - Period
  - Height
  - Wavelength
  - Direction
  - Steepness
  - "Groupiness"
- Bioluminescence
- Lunar phase/illumination
- Cloud cover
- Cloud ceiling
- Fog
- Aerosols
- Horizontal visibility
- Radio-frequency propagation
- Atmospheric refraction
- EM-EO ducting
- Precipitation

Submarine and Antisubmarine Warfare

Whereas submarine warfare remains a significant component of U.S. naval doctrine, threats to U.S. naval assets from enemy submarines have greatly diminished since the demise of the Soviet Union in the late 1980s and early 1990s. Nonetheless, ASW skills perfected by U.S. naval forces throughout the Cold War era are necessary elements of the capability to conduct multi mission operations. Submarines still serve as stealthy platforms to conduct covert reconnaissance or launch conventional cruise missiles and to stand guard as a nuclear deterrent force. However, the nature of submarine warfare and ASW in the littoral regions of the world brings with it a set of environmental parameters to which naval METOC functions are unaccustomed. These include the greater degree of spatial and temporal environmental variations in the littoral zone and their effects on the performance of submarine and antisubmarine sensors. Table B-9 provides some indication of the environmental parameters of importance to submarine warfare and ASW.

TABLE B-9 Environmental Parameters Impacting Submarine and Antisubmarine Warfare

- Winds at sea
- Waves/sea state (period)
    o Height
    o Wavelength
    o Direction
    o Steepness
    o "Groupiness"
    o Surface roughness
- Bioluminescence
- Ocean fronts
- Shelf currents
- Shelf bathymetry
- Seafloor acoustic properties
- Salinity variations
- Ocean temperature variations (lateral and vertical)
- Bubbles in the water column
- Ocean acoustic ducting
- Littoral zone ocean climatologies
- Seafloor type
- Acoustic transmission losses
- Subseabed geoacoustics

Appendix C

Environmental Science and Technology Programs

In considering the environmental information systems currently used by the U.S. Naval Forces, the committee has placed a great deal of interest and emphasis on the Navy's meteorological and oceanographic (METOC) community, which organizationally, falls under the Office of the Oceanographer of the Navy (N096). However environmental information plays a key role in programs and research initiatives, both basic and applied in other components of the Navy. These programs collect, analyze, and in some instances archive large datasets that may be of value to the METOC operations and METOC customers.

OFFICE OF NAVAL RESEARCH

The Office of Naval Research (ONR) sponsors science and technology in support of the U.S. Navy and Marine Corps. Founded in 1946, ONR today funds work at more than 450 universities, laboratories, and other organizations. The mission of the ONR is to maintain a close relationship with the research and development community to support long-range research, foster future discovery of technologies, and mature next generations of researchers for the future of the Navy and Marine Corps.

Department of Ocean, Atmosphere and Space (OAS)   (Code 32)

At ONR, OAS manages several science and technology programs in basic and applied research that provide environmental information in METOC to support the Navy and Marine Corps. The department consists of two large divisions—the Sensing and Systems S&T Division and the Processes and Prediction

S&T Division. These divisions provide multidisciplinary programs in naval environments, undersea warfare, and related subjects. The department also includes the Naval Space S & T Program Office, the central point of contact for the Department of the Navy's (DON) space science and technology activities. The department focuses its S&T programs in the areas of:

- *Battlespace Environments (BSE)*: Observing, modeling, and predicting both small- and large-scale processes in the air/ocean/shore environments. It contains the traditional oceanographic and meteorological disciplines and encompasses: Environmental Processes, Sensors/Data, Model Development, Data Assimilation and Information Exploitation, and Validation Studies.
- *Anti-Submarine Warfare (ASW)*: Detecting, localizing, and classifying submarines with active and passive acoustics as well as nonacoustic means. These are enhanced by automated data fusion and coupling with environmental understanding and modeling. Encompasses: Cooperative ASW, Wide-Area ASW Surveillance, and Battlegroup ASW Defense. This investment area includes the Littoral ASW Future Naval Capability effort.
- *Mine Warfare (MIW)*: Detecting, localizing, identifying, and neutralizing mines in both the ocean and littoral environment and improving offensive mining capabilities. Also includes Naval Special Warfare/Explosive Ordnance Disposal. Encompasses: Organic Minehunting (Sensing/Processing), Mine/Obstacle Neutralization, Sweeping/Jamming, Mining, and Advanced Force Operations. This investment area includes the Organic Mine Countermeasures future naval capabilities effort.
- *Maritime Intelligence, Surveillance, and Reconnaissance and Space Exploitation (ISR)*: Providing maritime situational awareness through development and exploitation of remote sensing and space capabilities. Encompasses: Remote/Space Sensing Processes, Space/Airborne Sensor Development, and Sensor Exploitation and Demonstration.

The Sensing and Systems S&T Division (Code 321)

The Sensing and Systems Division of the Ocean, Atmosphere, and Space Department conducts an extensive program of scientific inquiry and technology development in:

- Ocean Acoustics
- Remote Sensing and Space
- Sensing Information Dominance
- Coastal Dynamics
- Sensors, Sources, and Arrays
- Ocean Engineering and Marine Systems
- Undersea Signal Processing

The division's interests directly relate to Navy and Marine Corps operations, including undersea, expeditionary, and special warfare in littoral environments. In addition, the division manages the operation and maintenance of Navy research facilities, research ships, and other platforms.

Ocean Acoustics

The Ocean Acoustics Program supports research that addresses an understanding of the physics of the generation, propagation, and scattering of narrowband and broadband acoustic (and elastic) waves in the temporally and spatially varying ocean environment. Though research in acoustical signal processing is no longer a significant thrust of this program, ONR supports such research in collaboration with the Undersea Signal Processing Program, which also manages development and demonstration efforts. Research that uses acoustics solely as a tool to study other environmental processes should be proposed to the appropriate environmental programs (Geology and Geophysics, Physical Oceanography, Ocean Biology, etc.). However, ONR may jointly support the investigation of an acoustical tool or inverse method to probe some aspect of the environment, so long as its development is incomplete and relies on a better understanding of the relevant propagation and scattering mechanisms. In this case the investigator could also seek coordination of support from the appropriate environmental program officer. The Ocean Acoustics Program contains three primary thrusts and a "miscellaneous" thrust to capture topics of interest that clearly do not fit within any of the primary thrust areas. Brief descriptions of the three major thrusts are presented below.

- Shallow-Water Acoustics: the goal of this thrust is to understand the propagation and scattering of low-frequency (10 Hz to a few kHz) acoustic energy within the shallow-water ocean environment. Areas of research include investigations of the dominant shallow-water scattering mechanisms, the conversion of seafloor-incident acoustic energy into elastic body waves and interface waves, and acoustic propagation through linear and nonlinear internal waves.
- High-Frequency Acoustics: the goal of this thrust is to understand the interaction of high-frequency (few kHz to thousand kHz) sound with the ocean environment. Components of this thrust include the propagation of sound through an intervening turbulent or stochastic medium; scattering from rough surfaces, biologics, and bubbles; and penetration/propagation within the porous seafloor.
- Long-Range Propagation: the goal of this thrust is to understand the behavior of sound as it propagates over very long ranges (several hundred kilometers to several thousand kilometers) in the ocean. The main area of interest is understanding the effect of ocean internal waves on transmitted broadband acoustic signals.

Remote Sensing and Space

This program investigates physical and chemical processes that govern active and passive electromagnetic spectrum scattering from the Earth's surface and propagation through the upper atmosphere and the near-space environment. Of particular interest for surface effects are short water wave roughness modulation mechanisms, surfactant effects, intermittency in wave breaking, and non-linear water waves. Research is directed toward improving the knowledge base for development of mechanistic electro-optical/electromagnetic (EO/EM) clutter models and automatic target recognition and to investigate techniques that invert sensor information for the development of algorithms for assimilation into environmental models. Additional interests include electromagnetic scattering theory, microwave properties, scattering surface characterization, and wave and flux modulation mechanisms. Space research interests include improved specification of the global ionosphere and studies of ionospheric irregularities that impact radio frequency propagation at all frequencies up to and including those used by the global positioning system (GPS) system. Investigations of space weather phenomena are directed toward improved understanding and forecast of solar, heliospheric, and magnetospheric disturbances that produce C4I outages and destroy or degrade Naval space assets. Investigations of upper-atmospheric composition and dynamics are supported to improve specification of satellite drag and other space applications. Additional research interests include precise time and time interval, earth orientation, and astrometry for autonomous navigation and synchronization of Naval systems.

Sensing–Information Dominance

The Sensing–Information Dominance Program ensures that the Navy has the ability to form, interpret, forecast, and act on a complete tactical and environmental picture of the littoral undersea warfare (USW) and mine warfare (MIW) missions. Program investigators explore environmental effects on nonacoustic multisource data correlation and information display; automated reasoning under uncertainty; real-time multimedia databases; and large-scale, nonhardwired, high-throughput mobile computer networks. The program is involved in the use of nonacoustic undersea sensors and their related signal processing; science and technology for environmental and target data fusion for undersea warfare, mine warfare, expeditionary warfare, and worldwide ship tracking. Involvement in these areas aims to achieve robust, effective undersea surveillance in littoral regions; automatically form a seamless picture of the tactical situation, including its environment; and accurately assess the meaning of that evolving picture to enhance the performance of on-scene decision makers. Additional interests include: nonacoustic distributed system components, Ultra-Low Power light detection and ranging technology, superconducting quantum interference device radiometers, platform Anti-Submarine Warfare data fusion, airborne and shipborne periscope detection radar, wake detection, and the effects of surface ships on clouds. The Sensing Information Dominance Program contains five primary thrust areas:

- Non-Acoustic Mine and USW—Demonstrates the feasibility of non-acoustic technology and systems as an adjunct or alternative to acoustic technology for MIW and USW.
- MUSW Data Fusion—Develops basic understanding of data fusion methods and demonstrate data fusion technology and system feasibility applied to USW, MIW and METOC.
- Deployable Autonomous Distributed Systems—Supports the process required to surveil and control undersea battlespace through association, correlation and combination of information from multiple sensors to establish and maintain situation perception over the undersea battlespace and support attack on threat targets.
- Mining—Demonstrates feasibility of advanced mining technology and systems.
- Maritime ISR—Furthers the development of technology to automatically develop complete awareness of the littoral maritime situation long before, leading up to, during, and after military engagement.

Coastal Dynamics

The Coastal Dynamics Program includes aspects of the fluid and sediment mechanics of the coastal ocean. At present, two research areas are emphasized: (1) Nearshore Processes: fluid dynamics, fluid-sediment interactions, and the resulting morphological response of the nearshore, where waves begin to break because of shoaling; and (2) Surface Waves: the fluid mechanics of coastal surface waves and methods for improved prediction. There are collaborations with other programs to address issues such as coastal meteorology, littoral remote sensing, ocean models, and mine burial and migration. Emphasis is placed on model-driven experiments, where hypothesis and instrument locations are developed with models and subsequently evaluated with field studies. Sufficient insight has been gained in most Coastal Dynamics thrusts that the effects of three-dimensionality can no longer be ignored. Therefore, interest is also placed on the use of remote sensing and other techniques that can be used to augment and place in context, limited or minimal, in situ field deployments.

Sensors, Sources, and Arrays

This program conducts multidisciplinary science and technology development in all aspects of acoustic source and sensor systems for Navy surface ship, submarine, aircraft, or fixed ocean applications. These systems may be carried as onboard equipment or deployed and operated autonomously or remotely as mobile or fixed equipment. This program provides the next-generation acoustic source and sensor technology for the tactical and surveillance undersea warfare missions of the Navy. Project scopes may range from component-level research and development to system-level technology demonstrations. Areas of interest in undersea acoustic sensor technology that use environmental information include:

- Affordable technology
- Autonomous sensors with in-sensor signal processing
- Deployable sensor components, technology, and Concepts of Operations
- Energy storage technology, especially high-power/high-energy technology
- Environmental adaptation for acoustic sensors
- Environmental sensors in support of acoustic systems
- High-power, high-efficiency, low-cost, low-weight/volume transduction materials or designs
- Innovative sensor delivery and deployment concepts
- Innovative towed and hull system concepts for sensors
- Innovative towed system components and technologies
- Optical fiber cabled sensor systems
- Acoustic surveillance sensors and sources
- Tactical acoustic sensors and sources
- Volume efficient power components

Ocean Engineering and Marine Systems

The basic research component of the Ocean Engineering and Marine Systems S&T Program seeks to provide fundamental knowledge in interdisciplinary areas required for the development of innovative concepts for rapidly deployable, unmanned Marine Platform Systems in support of a broader more effective Naval presence at sea and in the littorals. Currently the major emphasis is to develop improved understanding and correspondingly improved modeling capability for nonlinear dynamics of flow-structure interaction. Primary research topics include wave and current loading mechanisms, structural response mechanisms, and coupled fluid-structure interaction. In addition to providing a fundamental knowledge framework for advanced engineering methodologies, this component seeks to accelerate transition of basic research developments into advanced marine platform systems with applications to unmanned surveillance and monitoring systems for the Naval forces and at-sea experimental capabilities for the ocean sciences community. In support of this objective, the program encourages cooperative research and technology development efforts combining basic and applied research investigators across ONR and other relevant Federally supported programs. The Ocean Engineering and Marine Systems S&T Program supports applied research and advanced technology development in support of the Organic Mine Warfare

Future Naval Capability and in particular in support of Naval Special Warfare (NSW), Explosive Ordnance Disposal (EOD), and U.S. Marine Corps Amphibious Landing Force operations. The major research and development thrusts in these components are:

- Sensor Technology (in support of diver and EOD operations)
- Surf and Beach Zone Clearance Technology (in support of in-stride breaching of mines and obstacles in advance of amphibious landing forces)
- Mission Support Technology (to improve the mobility, endurance, and effectiveness of NSW and EOD personnel)
- Applications of Autonomous Platform Systems to Mine Warfare (in order to minimize human involvement in mine field operations)

Undersea Signal Processing

The goal of the Undersea Signal Processing Program is to develop sonar signal processing algorithms that detect, identify, and locate quiet submarines, emphasizing quiet diesel-electric submarines operating in shallow water environments - and incoming threat weapons (torpedoes). To this end, ONR funds basic and applied research projects executed at universities, federal laboratories, and industry. The department's work is organized into the following thrust areas:

- Active sonar signal processing
- Passive sonar signal processing
- Fundamental research initiatives in signal processing

Active Sonar Signal Processing

The Active Sonar Signal Processing Program deals with environmental information related to operating active sonar systems in acoustically complex shallow water environments. These include:

- Detecting, classifying, and locating slow-speed diesel-electric submarines in shallow water
- Signal processing for autonomous active sonar systems
- Innovative signal processing algorithms for nonstationary clutter cancellation
- Multistatic active sonar system design
- Broadband signal processing algorithms and broadband target strength characterization
- Environmentally adaptive active sonar systems
- Detecting, classifying, and locating incoming threat weapons (torpedoes) in shallow water

Passive Sonar Signal Processing

The Passive Sonar Signal Processing Program's primary objective is to develop signal processing algorithms that detect, classify, and locate modern diesel-electric and nuclear submarines at tactically useful ranges in shallow water environments. These include:

• Innovative detection and classification algorithms that exploit nontraditional acoustic signatures and increase initial detection ranges
• Algorithms that operate effectively in nonstationary signal and noise fields
• Computationally efficient passive ranging techniques that can be easily integrated into conventional passive sonar processing strings
• Computationally efficient depth estimation techniques that can be easily integrated into conventional passive sonar processing strings
• Signal processing for autonomous passive sonar systems
• Environmentally adaptive passive sonar systems

Fundamental Research Initiatives Program

The program's goal is to broaden the Navy's science and technology base by conducting basic research in statistical and physics-based signal processing. Examples of research topics of interest include:

• Fundamental advances in detection and estimation using first principles of statistical decision and estimation theory
• Broadband signal processing algorithms and target-strength characterization
• Biologically-inspired underwater acoustic signal processing
• Feature sets for active sonar classification algorithms that can be used to distinguish returns produced by submarines and returns produced by other means
• Signal processing in underwater media characterized by time-varying multipath and random fluctuations
• Robust signal models that incorporate stochastic effects of underwater media
• The interdependence of array geometry, beam patterns, and the performance of reduced degree-of-freedom adaptive beamformers
• The performance of adaptive algorithms in cluttered environments, including a study of the adaptive degrees of freedom required to cancel discrete interference sources

The Processes and Prediction S&T Division (Code 322)

The Processes and Prediction Division of the Ocean, Atmosphere, and Space Department concentrates on improving the Navy and Marine Corps' understanding of environmental evolution, the assimilation of data, and the limits of predictability. It plans, fosters, and encourages an extensive program of scientific inquiry and technological development in fields ranging from environmental optics to high-latitude dynamics. Fields of special interest to the division include:

- Environmental Optics
- Physical Oceanography
- Biological and Chemical Oceanography
- Ocean Modeling and Prediction
- Marine Geosciences
- High Latitude Dynamics
- Marine Meteorology and Atmospheric Effects

Environmental Optics

The general goal of the Environmental Optics Program is to further understanding of how light interacts with the ocean, including the ocean boundaries (the sea surface and the ocean floor) and the atmosphere within tens of meters of the ocean surface.

Core Program

Funded basic research usually falls into one or more of the following categories:

- Radiative Transfer Modeling—developing and testing state-of-the-art numerical models of radiance propagation within the ocean.
- Instrument Development—developing the devices and techniques required to measure the inherent optical properties of ocean water and the ocean floor.
- Optical Process Studies—quantifying the interactions of light in the ocean with physical, biological, and chemical ocean processes.
- Coastal Remote Sensing—quantitative assessment of in-water inherent optical properties, bathymetry, and/or bottom type from high spectral- (primarily in the visible) and spatial-resolution aircraft or satellite data. The products of these basic research thrusts generally support the development or application of ocean prediction models, new ocean remote sensing systems, and associated image analysis algorithms.

Physical Oceanography

This program supports process-oriented and hypothesis-driven science and technology in the area of physical oceanography and in the applications of oceanic research tools and techniques for fleet usage. It should be noted that there are three other ONR teams involved with science and technology in the area of physical oceanography. These are Coastal Dynamics, Ocean Modeling and Prediction, and High Latitude. Strong interaction is encouraged with these programs. In response to post-Cold War U.S. Naval strategy and tactics, increased emphasis is given to the littoral, defined here as the oceanographic region encompassing the continental shelf and slope and the adjacent deep water. The approach is based principally on field observations with theory and modeling expected to be closely integrated. The following is a listing of the thrust areas with a brief description of emphasis:

- Air-sea interaction
Emphasis: impact on mixed-layer dynamics for improved parameterizations in oceanic and atmospheric models
- Internal waves/turbulence
Emphasis: development of a global littoral model
- Marginal seas/straits
Emphasis: identify key physics, develop archetypes, generalize results
- Shelf/slope dynamics
Emphasis: improved parameterizations for slope/shelf modeling. Examine coupling to deep water processes in particular boundary currents.
- Open ocean
Emphasis: emphasize meso/submesoscale spatial variability and time scales of weeks and smaller. Priority on upper-ocean processes. ONR encourages linkages to major international programs of global observations.
- Fleet meteorological and oceanographic (METOC) support
Emphasis: develop tools, techniques, and observations of direct impact to activities at sea.

Interdisciplinary Scientific Research

There is also a strong emphasis on interdisciplinary research with Biological/Chemical Oceanography, Ocean Acoustics, Environmental Optics, Marine Geology and Geophysics, and Marine Meteorology and Atmospheric Effects in research areas considered to be of high impact to the Navy and Marine Corps. An overarching objective of the program is to foster transition of research products such as numerical and theoretical models, analysis algorithms, in situ data, seagoing instrumentation and platforms into operational Naval systems. In addition, support is available for critical evaluation of the impact of the environment on fleet exercises that test and evaluate operational Naval systems.

Biological Oceanography

The goals of the biological oceanography program are:

- to enable prediction of the distribution, growth, and abundance of biota in the coastal ocean and shallow-water sediments, with the long-term goal of understanding how biota affect the optical and acoustical properties of operational importance to the Navy;
- to develop new instrumentation to sample and observe biological processes and phenomena; and
- to model and evaluate models of coupled bio-physical, bio-optical, and bio-acoustical processes in the coastal ocean and shallow-water sediments

Chemical Oceanography

The goals of the chemical oceanography program are:

- to enable prediction of chemical distributions and speciation in marine environments, especially as they relate to optical properties of seawater and interact with biota to influence optical and acoustical properties of seawater and sediments and
- to develop novel in situ chemical sensors to detect key chemical species rapidly, accurately, and at low detection threshold.

Specific investigators in chemical oceanography study the occurrence, production, and transformations of colored dissolved organic matter in the coastal ocean, air-sea gas exchange processes, aerosol chemical dynamics, chemistry of trace elements in the upper ocean, and nutrient dynamics.

Ocean Modeling and Prediction

This program seeks to develop accurate representations of the ocean system as it evolves in time and space. Underlying fundamentals include ocean field estimation, scale interaction and boundary interaction which are applied toward nowcast and forecast skill, subgrid-scale parameterization, ocean-atmosphere and ocean-bottom coupling and nested domains. The system includes acoustic and electromagnetic propagation models linked to hydrodynamic models. The goal of enhanced predictability is achieved through research on better dynamical formulations, improved numerical methods, and optimal data assimilation through adaptive sampling. Basic and applied research is pursued jointly with objectives to improve strategic and tactical decisions with environmental information and to motivate new understanding by operational experience.

APPENDIX C   183

Marine Geosciences

The Marine Geosciences program emphasizes studies related to the continental terrace (shelf and slope), and its composition, morphology, structure, and geological/ oceanographic processes. Its program priorities are focused by the Future Naval Capabilities (FNCs) initiative to develop stronger links among the Navy's requirements, acquisitions, and science and technology communities. Two FNCs of particular relevance to this program are Organic Mine Countermeasures (MCM) in very shallow water, surf, or beach zones and littoral ASW, aimed at sufficiently characterizing the battlespace environment to allow safe detection and monitoring of targets in shallow, nearshore environments. The overall program objective is to increase understanding of mechanisms that control structure, history, and dynamics of those geological features that in turn affect sound propagation, seafloor instability, bathymetry, and electromagnetic and optic transmissions in the water column and ocean bottom. These research areas of interest have been addressed in large part by the ONR Strata Formation on Margins (STRATAFORM) program, which is now winding down. STRATAFORM's overwhelming success has prompted initiation of a follow-on program, in collaboration with colleagues from European Union countries. The new EuroSTRATAFORM program will investigate relationships between active sediment dynamics on the continental shelf, cross-shelf transport and accumulation of sediment, and the preserved stratigraphic record. A key goal will be to test some of the new paradigms that have emerged from STRATAFORM and develop models that are capable of predicting stratigraphic patterns on a variety of continental margins. Three specific program thrust areas are outlined briefly below:

- *Continental Terraces.* Relate shelf sediment dynamics and the development of lithostratigraphy. Understand slope geological processes and resultant geomorphology. Interpret high-resolution character of stratigraphic sequences in shallow water resulting from shelf and slope sedimentation.
- *Sediment Processes.* Develop a fundamental understanding of the processes responsible for sediment motion, transport, bedform formation, and modification, and the development of seafloor stratigraphy particularly as these parameters affect Navy systems operating in shallow water.
- *Electromagnetics.* Various applied studies in electromagnetics, including instrument design and development, electromagnetic imaging of objects on the seafloor and the electromagnetic character of shallow water (continental shelf) environments.

High-Latitude Dynamics

This program investigates processes, primarily of a physical, biological, or chemical nature, that are active in the polar oceans. The emphasis is on the continental margins, including marginal seas and the adjacent slopes. The overarching program goal is to support the ongoing development of environmental models capable of supporting future fleet activities in the polar regions. Contributing objectives are to improve the Navy's understanding of ice mechanics and dynamics, air-sea-ice exchange processes, cross-shelf transport mechanisms, and turbulent mixing processes as they influence both upper-ocean mixing and deep convection. These program goals are addressed through individually funded research projects and through participation in coordinated, interagency research initiatives. Individual projects address diverse issues, including upper ocean turbulence, multiyear time series current measurements, ambient noise modeling and development of innovative systems such as Autonomous Underwater Vehicles (AUVs) and acoustic detection of interannual changes. The interagency Scientific Ice Experiment (SCICEX) program has concluded its field operations and is now in its final analysis phase. The joint ONR/National Science Foundation (NSF) Surface Heat Budget of the Arctic Ocean (SHEBA) experiment has concluded a successful field effort and is now entering its analysis phase. The joint ONR/NSF Western Arctic Shelf-Basin Interaction (SBI) study is in its pilot phase and will shortly enter its primary field phase. (Proposals for SBI are to be submitted in response to a forthcoming NSF/ONR Announcement of Opportunity.) The interagency international program for the Study of Environmental Arctic Change (SEARCH) is in the developmental phase. A new version of the Polar Ice Prediction System (PIPS) is under development for the Navy/National Ice Center jointly with the ONR Ocean Modeling Program.

Marine Meteorology and Atmospheric Effects

This program sponsors integrated basic, applied, and developmental research with emphasis on improving the modeling and prediction of environmental parameters critical to Navy and Marine Corps platform, sensor, and weapons performance. The program includes research and development leading to enhanced environmental support for operations, training, mission planning, and systems development, and focuses on addressing Navy and Marine Corps real-time, high-resolution environmental requirements to support tactical sensors and operations in littoral zones worldwide. Topics of interest include marine boundary layer processes including aerosols, marine convective, and nonconvective clouds; mesoscale coastal phenomena such as coastally-trapped disturbances; data assimilation incorporating high data rate, asynchronous sensors (radar, lidar, etc.); global, mesoscale, and on-scene modeling focusing on the marine atmosphere and/or coastal zone; atmospheric predictability; environmental effects on electromagnetic and electrooptic propagation, and western and southern Pacific tropical cyclone behavior and evolution in motion and structure.

THE NAVAL SYSTEMS COMMAND

The Systems Commands provide for and meets those material support needs of the DON that are within the assigned "material support" responsibility of each command. This general responsibility includes the research, design, development, logistics planning, testing, technical evaluation, acquisition, procurement, contracting, production, construction, manufacture, inspection, fitting out, supply, maintenance, alteration, conversion, repair, overhaul, modification, advance base outfitting, safeguarding, distribution, and disposal of naval material. In addition, individual Systems Commands are tasked to perform control, coordination, or service functions as designated Lead Systems Commands for particular programs or functions. Environmental information from these commands plays a key role in enabling the Navy to operate its ships, aircraft, and various weapons and sensor systems in a complex and changing environment. The purpose of this section is to discuss these entities to show how they fit into the Navy's use of environmental information.

Naval Sea Systems Command

The Naval Sea Systems Command (NAVSEA) designs, develops, builds, and maintains the U.S. Naval Fleet, ships, shipboard weapons, and combat systems. It is headquartered in Washington, D.C., and Arlington, Virginia, NAVSEA, and is the U.S. Navy's ship systems program manager, engineer, and technical authority. The largest of the Navy's five systems commands, NAVSEA manages approximately 130 acquisition programs and provides engineering, technical authority, and logistics support to the fleet via its headquarters operations; six affiliated Program Executive Offices (PEOs); four Naval shipyards; nine Supervisors of Shipbuilding, Conversion, and Repair; two technical warfare centers; and numerous subordinate organizations and offices. NAVSEA also administers more than 1,400 foreign military sales cases worth $16.7 billion, involving 80 countries and four NATO organizations. The six affiliated PEOs, managed by NAVSEA are Aircraft Carriers, Surface Strike, Expeditionary Warfare, Mine and Undersea Warfare, Submarines, and Theater Surface Combatants. The PEOs are responsible for all aspects of life-cycle management for these programs. NAVSEA provides the PEOs with total ship system engineering; establishes and coordinates technical policy, directives, and procedures governing ship and ship system technical requirements; and provides integrated logistics support. NAVSEA is the largest of the Navy's five systems commands. It accounts for nearly one-fifth of the Navy's budget (approximately $20 billion) and manages more than 130 acquisition programs, which are assigned to the six PEOs. The Command consists of a Headquarters organization at the Washington Navy Yard and a variety of technical and industrial organizations located throughout the country. The Command's major organizations outside headquarters generally are grouped into two technical centers, an ordnance center, the public naval shipyards, and the Supervisors of Shipbuilding, Conversion, and Repair (SUPSHIPs) that oversee the Command's new ship construction and in-service ship repair efforts. The Naval Sea Systems Command provides environmental engineering and technical support to:

- Ships, submersibles, other sea platforms, and craft
- Shipboard combat systems, including sensors, tactical data systems, surveillance and fire control radars, sonars, computers, guns, launchers, ammunition, guided missiles, mines, and torpedoes
- Shipborne components, including nuclear and nonnuclear propulsion, electrical generating equipment, auxiliary power generating and distribution systems, interior communications, navigation equipment, deck machinery, weapons and cargo handling, stowage, and damage control systems
- Diving and salvaging equipment
- Explosive ordnance disposal and explosive safety
- Ship systems integration
- Serves as lead SYSCOM for logistics research and development
- Weapons systems program support
- Materials-handling equipment not otherwise assigned
- Special clothing not otherwise assigned
- Automation of Navy technical data
- Naval material for which responsibility is not otherwise assigned

Naval Air Systems Command

The Naval Aviation Systems Command (NAVAIR) is recognized for developing, acquiring, and providing environmental support to maritime aeronautical systems that can be operated and sustained at sea. NAVSEA's philosophy enables the entire organization to operate seamlessly—to carry out the national defense strategy. Ensuring the highest readiness priorities while providing the best value to the fleet is a top priority at NAVAIR. Future readiness depends largely on technological superiority and integration of joint operating forces in joint battle spaces. NAVAIR is comprised of six organizations. Naval Air Systems Command (NAVAIR); Naval Inventory Control Point (NAVICP); Program Executive Office, Air Anti-Submarine Warfare, Assault, and Special Mission Programs PEO(A); Program Executive Office, Tactical Aircraft Programs PEO(T); Program Executive Office, Strike Weapons and Unmanned Aviation PEO(W); and Program Executive Office, Joint Strike Fighter PEO(JSF). Within the Naval Air Systems Command the Naval Air Warfare Center Training Systems Division (NAWCTSD) has a long history of technology transfer to both the public and private sectors. NAWCTSD is involved with the local school system, NASA

Kennedy Space Center, and the Federal Aviation Administration to share information and expertise. There are currently five Cooperative Research and Development Agreements (CRADAs). CRADAs provide for the transfer of technology developed in federal government laboratories to the private sector. By sharing Navy training research, the public will benefit in having improved education and training. The Navy also receives valuable information in the exchange of information and resources. NAVAIR has responsibilities in the following areas:

- Navy and Marine Corps aircraft systems and components (including fuels and lubricants)
- Air-launched weapons systems and components (excluding torpedoes and mines)
- Other airborne and air-launched systems and components such as electronics, underwater sound, catapults, aircraft/missile range and evaluation instrumentation, mine countermeasures, targets, logistical equipment, and training and support systems for the foregoing

Naval Underwater Warfare Center

The Naval Undersea Warfare Center (NUWC), officially established on January 2, 1992, is the Navy's full-spectrum research, development, test and evaluation, engineering, and fleet support center for submarines, autonomous underwater systems, and offensive and defensive weapons systems associated with undersea warfare. There are two major divisions of this Warfare Center—Division Newport located in Newport, Rhode Island, and Division Keyport located in Keyport, Washington. NUWC was formed by consolidating the Naval Underwater Systems Center, Newport, and the Naval Undersea Warfare Engineering Station, Keyport. In addition to its two main sites at Newport and Keyport, NUWC has several detachments geographically spread across North America: from Andros Island, Bahamas, to Lualualei, Hawaii, and from San Diego, California, to Nanoose, British Columbia. The Warfare Center seeks to provide the highest-quality technologies and services at the best value to ensure the Nation's continuing superiority in undersea warfare. Along with technical programs, NUWC has continued the tradition that began with its predecessor organizations—demonstrating exceptionally superior performance in the areas of technical management, planning, business management, and human resources. The Naval Undersea Warfare Center is the U.S. Navy's full-spectrum research, development, test and evaluation, engineering, and fleet support center for submarines, autonomous underwater systems, and offensive and defensive weapons systems associated with Undersea Warfare. NUWC has been assigned responsibility over the full spectrum of undersea weapons systems from Science and Technology to In-service Support—"Cradle to Grave." One of the primary objectives of NUWC is to establish and maintain synergy of operations and capabilities within the undersea warfare systems community. The NSWCDD mission is to provide research, development, test and evaluation, engineering, and fleet support for:

- Surface warfare
- Surface ship combat systems
- Ordnance
- Strategic systems
- Mines
- Amphibious warfare systems
- Mine countermeasures
- Special warfare systems

Space and Naval Warfare Systems Center

SPAWAR's mission is to provide the warfighter with knowledge superiority by developing, delivering, and maintaining effective, capable, and integrated command, control, communications, computer, intelligence, and surveillance systems. While its name and organizational structure have changed several times over the years, the basic mission of helping the Navy communicate and share critical information has not. SPAWAR provides information technology and space systems for today's Navy and Defense Department activities while planning and designing for the future. The Department of the Navy established the Naval Electronic Systems Command (NAELEX) on May 1, 1966, to provide the U.S. Navy and Marine Corps operating forces with the best Command, Control and Communications electronic systems. NAVELEX engineers, scientists, technicians and support employees worked to meet the demands of their mission. With the approach of the 21st century, the Navy Department was reevaluated to maximize its strengths, and a major re-organization took place. The Navy disestablished the Material Command, and in May 1985 NAVELEX became the Space and Naval Warfare Systems Command (SPAWAR)—an Echelon II Command under the Chief of Naval Operations. With the new name came new responsibilities. In addition to meeting the fleet's Command, Control and Communications requirements, emphasis was placed on Undersea Surveillance and Space Systems programs. SPAWAR became the Navy's Battle Force Architect—a new concept aimed at designing total systems for the forces instead of individual platforms and weapons. With the mission change, SPAWAR became manager of eight Navy laboratories and four university laboratories, as well as seven engineering centers geographically dispersed throughout the country. SPAWAR comprises five program directorates at its headquarters in San Diego, California, three systems centers—Charleston, Chesapeake, and San Diego—the SPAWAR Space Field Activity (SSFA) and the SPAWAR Information Technology Center. SPAWAR is one of five Navy Acquisition Commands. SPAWAR Systems Center, Chesapeake, is located in Chesapeake, Virginia, and provides software design, development, testing, training, delivery, and support operations. Branch offices in San Diego, California; Sigonella, Italy; and Yokosuka, Japan; provide sailors, Airmen, and Marines with comprehensive, life-cycle support of all products and services across the globe and around the clock. SPAWAR Systems Center, Chesapeake, designs, develops, delivers, and supports integrated information systems for the Navy and the U.S. Marine Corps. Logistical support for the operating forces is more complex today than ever before and keeping the Navy's ships, submarines, and aircraft in peak operating condition demands reliable and responsive business information systems. Located in San Diego, with detachments in Hawaii, Guam, and Japan, SPAWAR Systems Center, San Diego employs nearly 4,000 civilian and military personnel. The capabilities they develop allow the Navy's decision-makers and, increasingly, the joint services, to protect their own forces and carry out their operational missions. Information—financial, administrative, and statistical—is the lifeblood of the modern world. For the Navy's tactical commanders at sea, information can mean the difference between victory and defeat, life and death. SPAWAR Systems Center, San Diego, is responsible for developing technology that collects, transmits, processes, displays and, most critically, manages information essential to naval operations. The SPAWAR Space Field Activity (SSFA), located in Chantilly, Virginia, provides line management staffing of the National Reconnaissance Office. SSFA personnel coordinate naval space research, development, and acquisition activities between NRO and other space programs. The SSFA also provides naval space and warfare experience to develop superior and affordable space systems in support of national missions and joint, combined, and naval operations. The result of a merger in December 2000, the SPAWAR Information Technology Center is located in New Orleans, Louisiana. It provides high-quality information management and information technology products, services, and solutions to satisfy requirements of the DON, DOD, and other government agencies. SPAWAR provides material and environmental support to:

- Command/control/communications (C3) (platform to platform)
- Undersea and space surveillance (includes short communications)
- Marine Corps expeditionary and amphibious electronics
- Multiplatform electronic systems not otherwise assigned
- Intelligence and intelligence collection systems
- Space systems
- Cryptographic and cryptologic equipment

In addition, SPAWAR has DON-wide responsibility for warfighting architecture development and requirements integration among the total naval battle force; to provide similar material support for the Marine Corps; and to provide management of DON R&D Centers.

DEFENSE ADVANCED RESEARCH PROJECTS AGENCY

The Defense Advanced Research Projects Agency (DARPA) is the central research and development organization for the DOD. In this capacity it sponsors a great deal of research in all areas of the environment. It manages and directs selected basic and applied research and development projects for DOD and pursues research and technology where risk and payoff are both very high and where success may provide dramatic advances for traditional military roles and missions. The Defense Advanced Research Projects Agency (DARPA) mission is to develop imaginative, innovative, and often high-risk research ideas offering a significant technological impact will go well beyond the normal evolutionary developmental approaches and to pursue these ideas from the demonstration of technical feasibility through the development of prototype systems.

DARPA was established in 1958 as the first U.S. response to the Soviet launching of Sputnik. Since that time DARPA's mission has been to assure that the United States maintains a lead in applying state-of-the-art technology for military capabilities and to prevent technological surprise from her adversaries. The DARPA organization is as unique as its role, reporting directly to the Secretary of Defense and operating in coordination with, but completely independent of, the military research and development (R&D) establishment. Strong support from senior DOD management has always been essential since DARPA was designed to be an anathema to the conventional military and R&D structure and, in fact, to be a deliberate counterpoint to traditional thinking and approaches. Some of the more important founding characteristics are listed below. Over the years DARPA has continued to adhere to these founding principles:

- Small and flexible
- Flat organization
- Substantial autonomy and freedom from bureaucratic impediments
- Technical staff drawn from world-class scientists and engineers with representation from industry, universities, government laboratories and Federally Funded Research and Development Centers
- Technical staff assigned for 3-5 years and rotated to assure fresh thinking and perspective;
- Project based—all efforts are typically 3-5 years long with strong focus on end goals. Major technological challenges may be addressed over much longer times but only as a series of focused steps. The end of each project is the end. It may be that another project is started in the same technical area, perhaps with the same program manager and, to the outside world, this may be seen as a simple extension. For DARPA, though, it is a conscious weighing of the current opportunity and a completely fresh decision. The fact of prior investment is irrelevant.
- Necessary supporting personnel (technical, contracting, administrative) are "hired" on a temporary basis to provide complete flexibility to get into and

out of an area without the problems of sustaining the staff. This is by agreement with DOD or other governmental organizations (military R&D groups, NASA, NSF, etc.) and from System Engineering and Technical Assistance (SETA) contractors.
- Program Managers (the heart of DARPA) are selected to be technically outstanding and entrepreneurial. The best DARPA Program Managers have always been free-wheeling zealots in pursuit of their goals,
- Management is focused on good stewardship of taxpayer funds but imposes little else in terms of rules. Management's job is to enable the Program Managers to expand support.
- A complete acceptance of failure if the payoff of success was high enough.

DARPA Technical Offices

Advanced Technology Office (ATO)

ATO researches, demonstrates, and develops high payoff projects in maritime, communications, special operations, command and control, and information assurance and survivability mission areas. These projects support military operations throughout the spectrum of conflict. ATO adapts advanced technologies into military systems and also exploits emerging technologies for future programs. The ultimate goal is superior cost-effective systems the military can use to respond to new and emerging threats.

Defense Sciences Office (DSO)

The mission of the DSO is to vigorously pursue the most promising discoveries and innovations in science and engineering to create paradigm shifts in defense capabilities. DSO emphasizes programs in medical approaches to biological warfare defense, biology, materials, and advanced mathematics.

Information Awareness Office

This office develops and demonstrates information technologies and systems to counter asymmetric threats by achieving total information awareness useful for preemption, national security warning, and national security decision-making.

Information Technology Office

The Information Technology Office focuses on inventing the networking, computing, and software technologies vital to ensuring DOD military superiority.

Information Exploitation Office (IXO)

This office develops sensor and information system technology and systems with application to battlespace awareness, targeting, command and control, and the supporting infrastructure required to address land-based threats in a dynamic, closed-loop process. IXO leverages ongoing DARPA efforts in sensors, sensor exploitation, information management, and command and control and addresses systemic challenges associated with performing surface target interdiction in environments that require very high combat identification confidence and an associated low likelihood for inadvertent collateral damage.

Microsystems Technology Office

This office's mission focuses on the heterogeneous microchip-scale integration of electronics, photonics, and microelectromechanical systems. The high-risk/high-payoff technology is aimed at solving national-level problems of protection from biological, chemical, and information attack and to provide operational dominance for mobile distributed command and control, combined manned/unmanned warfare, and dynamic adaptive military planning and execution.

Special Projects Office (SPO)

The SPO focuses on developing systems solutions, along with the required enabling technologies, to counter current and emerging threats. In the area of current challenges, SPO is focused on affordable precision kill of movers, emitters, and concealed (including underground) targets. In the area of emerging threats, SPO focuses on active defenses against biological weapons, proliferated, low-cost/low-technology air vehicles and missiles, and GPS jamming. Supporting technologies include advanced sensors and radars, signal processing, and navigation and guidance systems.

Tactical Technology Office

The Tactical Technology Office engages in high-risk, high-payoff advanced military research, emphasizing the "system" and "subsystem" approach to the development of aeronautic, space, and land systems as well as embedded processors and control systems.

NOT-FOR-PROFIT ACTIVITIES SUPPORTING NAVAL ENVIRONMENTAL R&D

A great deal of environmental support for naval systems, ships, aircraft, and sensors is also done at non-for-profit activities supporting naval research and development. A brief synopsis of the mission of these installations is given below:

Marine Physical Laboratory, Scripps Institution of Oceanography

Mission: To generate knowledge about the ocean and its boundaries and application of this knowledge to the solution of Navy undersea problems.

Applied Research Laboratory, Pennsylvania State University

Mission: To (1) serve as the lead laboratory for research in the guidance and control of undersea weapons; (2) provide corporate memory and technical expertise in the area of advanced closed-cycle thermal propulsion systems for undersea weapons; and (3) provide expertise in the area of propulsion technology, hydrodynamics, and hydro acoustics for undersea vehicles and weapons.

Applied Research Laboratories, University of Texas at Austin

Mission: To (1) contribute to fundamental scientific advances in acoustics and electromagnetics; (2) help with exploitation of relevant research results; and (3) conduct RDT&E and field support for solution of Navy wartime problems in acoustics and electromagnetics for surface, subsurface, and space environments.

Applied Physics Laboratory, University of Washington

Mission: To conduct a university-based program of fundamental research, technology advancement, and engineering support emphasizing naval applications of ocean science, ocean acoustics, and engineering.

Applied Physics Laboratory, Johns Hopkins University

Mission: To provide essential engineering, research, development; test and evaluation capabilities in support of programs to improve the efficiency and assure the availability of current and future Navy strategic and tactical forces; and to conduct related scientific and technical programs on behalf of other military and civilian agencies of the government.

Systems Research Center, Virginia Polytechnic Institute and State University

Mission: To conduct research and development for computing support systems of interest to the Navy and other government agencies.

Center for Naval Analyses (CAN)

Mission: To conduct a continuing program of research, studies, and investigations that will provide information needed for DON management decisions addressing the development and application of naval capabilities, help the operating forces of DON improve their effectiveness, and develop operational data for use in force planning and force evaluation studies.

Appendix D

Acronyms

4D Cube	Four-Dimensional Cube
ADS	Advanced Deployable System
AMW	Amphibious Warfare
ANDES	Ambient Noise Directionality Estimation System
ASW	Antisubmarine Warfare
AT/FP	Anti-Terrorism/Force Protection
ATO	Air Tasking Order
AUV	Autonomous Underwater Vehicle
AWM	Amphibious Warfare
ARCI	Acoustic Rapid COTS (commercial-of-the-shelf) Insertion
BLUG	Bottom Loss Upgrade
BMDA3	Battlespace METOC Data Acquisition, Assimilation and Application
BSE	Battlespace Environments
C4I	Command Control Communication and Computer Intelligence
CEC	Cooperative Engagement Capability
CEP	Circular Error Probable
CINC	Commander in Chief
CNA	Center for Naval Analyses

CNMOC	Commander of Naval Meteorology and Oceanography Command
CNO	Chief of Naval Operations
CNR	Chief of Naval Research
CODE 32	Department of Ocean, Atmosphere and Space, ONR
CODE 321	Sensing and Systems S&T Division, ONR
CODE 322	Processes and Prediction S&T Division, ONR
COE	Center of Expertise
COAMPS	Coupled Oceanographic Atmospheric Mesoscale Prediction System
CONOPS	Concept of Operations
CRADA	Cooperative Research and Development Agreements
CROP	Common Relevant Operating Picture
CTD	Conductivity, Temperature, Depth
CVBG	Carrier Battle Group
CVN	Carrier, Nuclear
DAMPS	Distributed Atmospheric Mesoscale Prediction System
DANES	Directional Ambient Noise Estimation System
DARPA	Defense Advanced Research Projects Agency
DEAD	Destruction of Enemy Air Defense
DIFAR	Digital Frequency and Ranging
DOD	Department of Defense
DON	Department of the Navy
DRI	Defense Research Initiative
DSO	Defense Sciences Office
DTRA	Defense Threat Reduction Agency
EBO	Effects-Based Operations
EM	Electromagnetic
EMPRA	Embarkation, Movement, Planning, Rehearsal, and Assault
ENSO	El Nino Southern Oscillation
EO	Electro-optical
EOD	Explosive Ordnance Disposal
ESG	Expeditionary Sensor Grid
ETOPO5	Earth Topography—5 minutes
EXW	Expeditionary Warfare
EW	Electronic Warfare
FNCs	Future Naval Capabilities
FNMOC	Fleet Numerical Meteorology and Oceanography Command

GDEM	Generalized Digital Environmental Model
GGI&S	Global Geospatial Information and Services
GPS	Global Positioning System
HLD/HLS	Homeland Defense/Homeland Security
HARM	High-Speed Anti-radiation Missile
Hz	Hertz
IRC	Internet Relay Chat
ISR	Intelligence, Surveillance, Reconnaissance
IT-21	Information Technology 21st Century
IWAR	Integrated Warfare Area
JDAM	Joint Direct Attack Munition
JEM	Joint Effects Model
kHz	Kilohertz
LFA	Low-Frequency Acoustic
LFAA	Low-Frequency Active Adjunct
LMRS	Long-term Mine Reconnaissance System
LOTS	Logistics-Over-The-Shore
MAP	Maritime Air Patrol
MCM	Mine Countermeasures
MEMS	Microelectromechanical systems
METMF(R)	Meteorological Mobile Facility (Replacement)
METOC	Meteorology and Oceanography
MIW	Mine Warfare
MMA	Multi-Mission Aircraft
MMS	Multi-Mission Ship
MODAS	Modular Ocean Data Assimilation System
MUSW	Mine and Undersea Warfare
N096	Oceanographer of the Navy
NAELEX	Naval Electronic Systems Command
NASA	National Aeronautics and Space Administration
NATO	North Atlantic Treaty Organization
NAVAIR	Naval Aviation Systems Command
NAVELEX	Naval Electronic Systems Command
NAVICP	Naval Inventory Control Point
NAVOCEANO	Naval Oceanographic Office
NAVSEA	Naval Sea Systems Commands

NAVSPECWARCOM	Naval Special Warfare Command
NAWCTSD	Naval Air Warfare Center Training Systems Division
NCMI	Navy-Marine Corps Internet
NCO	Network-Centric Operations
NCW	Network-Centric Warfare
NFN	Naval Fires Network
NOAA	National Oceanic and Atmospheric Administration
NRC	National Research Council
NRO	National Reconnaissance Office
NSF	National Science Foundation
NRL	Naval Research Laboratory
NSWCDD	Naval Surface Warfare Center, Dahlgren Divison
NUWC	Naval Undersea Warfare Center
NWDC	Navy Warfare Development Command
NWP	Numerical Weather Prediction
NSW	Naval Special Warfare
OAS	Ocean, Atmosphere and Space
OEC	Optimized Environmental Characterization
OEO	Other Expeditionary Operations
OMFTS	Objective Maneuver From The Sea1
ONR	Office of Naval Research
OPARS	Optimal Path Aircraft Routing Service
OSB	Ocean Studies Board
OTH	Over The Horizon
OTSR	Optimum Track Ship Routing Service
PCIMAT	Personal Computer Interactive Multisensor Analysis Training
PEOs	Program Executive Offices
PEO(A)	Program Executive Office, Air Anti-Submarine, Warfare, Assault, and Special Mission Programs
PEO(JSF)	Program Executive Office, Joint Strike Fighter
PEO(T)	Program Executive Office, Tactical Aircraft Programs
PEO(W)	Program Executive Office, Strike Weapons and Unmanned Aviation
PERMA	Planning, Embarkation, Movement, Rehearsal, Assault
PGMs	Precision-Guided Munitions
PIPS	Polar Ice Prediction System
PTA	Precise Time and Astrometry
PUMA	Precision Undersea Mapping
QDR 2001	Quadrennial Defense Review 2001

R&D	Research and Development
RDT&E	Research, Development, Test, and Evaluation
REA	Rapid Environmental Assessment
RMS	Remote Minehunting System
ROC	Relative Operating Characteristics
SABLE	Sonar Active Bottom Loss Estimate
SBI	Shelf-Basin Interaction study
SCICEX	Scientific Ice Formation
SDV	Swimmer Delivery Vehicle
SEAD	Suppression of Enemy Air Defense
SEARCH	Study of Environmental Artic Change
SETA	System Engineering and Technical Assistance
SFMPL	Submarine Fleet Mission Profile Library
SHEBA	Surface Heat Budget of the Artic Ocean
SIAP	Single Integrated Air Picture
SMOOS	Shipboard Meteorological and Oceanographic Observing System
SOSUS	Sound Surveillance System
SPAWAR	Space and Naval Systems Command
SPO	Special Projects Office
SSD	Ship Self Defense
SSFA	Space Field Activity
SSP	Sound Speed Profile
SST	Sea Surface Temperature
S&T	Science and Technology
STOIC	Special Tactical Oceanographic Information Chart
STOM	Ship To Objective Maneuver
STRATAFORM	Strata Formation on Margins
STW	Strike Warfare
SUPSHIPs	Supervisors of Shipbuilding, Conversion and Repair
SURTASS	Surveillance Towed Array System
SUW	Surface Warfare Superiority
SYSCOM	Systems Command
T-AGS(X)	Multi-Mission Ship
TCS	Time Critical Strike Operations
TDA	Tactical Decision Aid
TEDS	Tactical Environmental Data Server
TMD	Theater Missile Defense
TPFD	Time Phased Force Deployment Directive
TTS	Through The Sensor
TTST	Through-The-Sensor Technology

UAV	Unmanned Aerial Vehicle
USMC	United States Marine Corps
USNO	U.S. Naval Observatory
USV	Unmanned Surface Vehicle
USW	Undersea Warfare
UTC	Coordinated Universal Time
UUV	Unmanned Undersea Vehicle
VLADS	Vertical Line Acoustic DIFAR (Digital Frequency and Ranging)
VNE	Virtual Natural Environment
VSW	Very Shallow Water
WAA	Wide Aperture Array
WEN	Web-Enabled Navy
WGS-84	World Geodetic Standard-1984
WMD	Weapons of Mass Destruction
WME	Weapons of Mass Effect

Appendix E

Information Gathering Activities of the Committee on Environmental Information for Naval Use

March 12-13, 2001, Washington, D.C. (full committee). Presentations by the Office of the Oceanographer of the Navy, the Office of Naval Research, and the Naval Meteorology and Oceanography Command provided the committee with an overview of each organization and a sense of the basic components of the current METOC enterprise.

May 21-23, 2001, Norfolk Naval Base, Norfolk, Va. (full committee). Presentations by personnel from the Office of the Oceanographer of the Navy, the Naval Atlantic Meteorology and Oceanography Center (located at Norfolk Naval Base) and the Naval Meteorology and Oceanography Command were supplement by discussions with personnel attached to Cruiser-Destroyer Group 2, the Office of the Assistant Secretary of the Navy for Acquisition, Research, and Development, and the Office of Naval Research to provide the committee with an understanding of needs of the surface fleet and mine warfare community and efforts to address them.

July 20, 2001, Naval War College, Newport, R.I. (sub-group meeting). Discussion with personnel from the Naval War College (including then-president VADM Arthur K. Cebrowski) and the Naval Warfare Development Command explored aspects of network-centric warfare as it pertain METOC operational concepts and the use of METOC products in training exercises and wargames.

August 6, 2002, CNMOC, Stennis, Miss. (sub-group meeting). Discussions with personnel from the Naval Meteorology and Oceanography Command, the

Naval Oceanographic Office, and the Warfighter Support Center focused on the acquisition, management, and dissemination of geospatial data.

August 7-9, 2001, Stennis Space Center, Stennis, Miss. (full committee) Presentations by personnel from the Office of the Oceanographer of the Navy, the Naval Meteorology and Oceanography Command, the Naval Oceanographic Office, the Naval Ice Center, and the Joint Staff (N6) focused on daily demands for METOC products, efforts to quantify and minimize uncertainty in model products, and the implications of network-centric principles for METOC operations.

November 5-7, 2001, Naval Pacific Meteorology and Oceanography Center, Pearl Harbor, Hi. (sub-group meeting). Discussions with personnel at the Naval Pacific Meteorology and Oceanography Center, Pearl Harbor, focused on products generated by the Typhoon Warning Center and the use of classified imagery to generate METOC observations.

November 14-16, 2001, Naval Pacific Meteorology and Oceanography Center, San Diego, Cal., and Fleet Numerical Meteorology and Oceanography Center, Monterey, Cal. (full committee). Presentations by personnel from the Naval Pacific Meteorology and Oceanography Center, San Diego, focused on rapid environmental assessment and challenges for supporting various weapons systems (including the Naval Fires Network) and Naval Special Warfare operations. Presentations by personnel from the Fleet Numerical Meteorology and Oceanography Center, Monterey focused on meteorological modeling efforts including efforts to provide nested model products.

December 11, 2001, Naval War College, Newport, R.I. (sub-group meeting). Discussion with personnel form the Naval War College and the Naval Warfare Development Center focused on current efforts to support strike warfare and the use and value of tailored products provided by the Fleet Numerical Meteorology and Oceanography Center, Monterey.

March 18, 2002, OPNAV, Crystal City, Va. (sub-group meeting). Discussions with personnel from the Naval Oceanographic Office, Submarine Development Squadron 12, and Submarine Warfare Division (N77) focused on tailored environmental products and the design and functionality of tactical decision aids for undersea warfare.

August 12, 2002, Camp Pendelton, San Diego, Cal. (sub-group meeting). Discussions with the senior USMC METOC officer attached to I Marine Expeditionary Force focused on USMC-specific needs for environmental information and efforts and products intended to address them. The relationship between

USMC and U.S. Navy METOC was explored, as well as the practical relationships between intelligence, reconnaissance, and surveillance efforts and environmental data collection.

September 26, 2002, Defense Threat Reduction Agency, Alexandria, Va. (sub-group meeting). Discussions with personnel from the Defense Threat Reduction Agency focused on environmental data needs to evaluate the battlefield threat from weapons of mass destruction, the collection of environmental data in denied areas, and barriers to effective and timely sharing of environmental data from these denied areas.